William Bernhard Tegetmeier, Francis Orpen Morris

A Natural History of the Nests and Eggs of British Birds

Vol. II

William Bernhard Tegetmeier, Francis Orpen Morris

A Natural History of the Nests and Eggs of British Birds
Vol. II

ISBN/EAN: 9783744790932

Printed in Europe, USA, Canada, Australia, Japan

Cover: Foto ©berggeist007 / pixelio.de

More available books at **www.hansebooks.com**

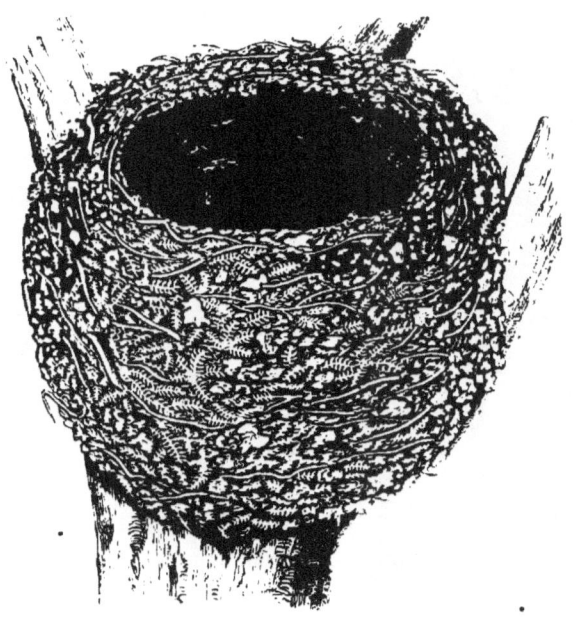

CHAFFINCH

A NATURAL HISTORY

OF THE

NESTS AND EGGS

OF

BRITISH BIRDS

BY THE

REV. F. O. MORRIS, B.A.

RECTOR OF NUNBURNHOLME, YORKSHIRE

FOURTH EDITION

REVISED AND CORRECTED BY

W. B. TEGETMEIER, F.Z.S.

MEMBER OF THE BRITISH ORNITHOLOGISTS' UNION

WITH TWO HUNDRED AND FORTY-EIGHT PLATES

CHIEFLY COLOURED BY HAND

IN THREE VOLUMES

VOLUME THE SECOND

LONDON

JOHN C. NIMMO

14 KING WILLIAM STREET, STRAND

MDCCCXCVI

Printed by BALLANTYNE, HANSON & CO.
At the Ballantyne Press

CONTENTS AND PLATES

VOLUME THE SECOND

CONTENTS AND PLATES

vii

NESTS AND EGGS

OF

BRITISH BIRDS

CHAFFINCH

SHILFA—SCOBBY—SHELLY—SKELLY—SHELL-APPLE—BEECH-
FINCH—TWINK—SPINK—PINK—TWEET.

PLATE LXXX.

Fringilla cælebs   .   .   .   .   Linnæus.

THE nest of the Chaffinch is truly a beautiful piece of
workmanship, compact and neat in the highest degree.
It is usually so well adapted to the colour of the place
where it is built, as to elude detection from any chance
passer-by—close scrutiny is required to discover it. It
is therefore variously made, according to the nature of the
materials at hand. Some are built of grasses, stalks of plants,
and small roots, compacted with the scales of bark and
wool, and lined with hair, with perhaps a few feathers, the
outside being entirely covered with tree moss and lichens,
taken from the tree itself in which it is placed; the assi-
milation being thus rendered complete. Others are without
any wool, its place being supplied with thistle-down, and

spider-webs. In fact the bird accommodates itself to circumstances, using such materials as are at hand. The upper edge of the nest is generally very neatly woven with slender fibres, and the width of the open part is often not more than an inch and a half, but usually an inch and three quarters; the whole is firmly fixed between the branches, to which some of its component parts are attached.

> "———the Shilfa's nest, that seems to be
> A portion of the sheltering tree."

In the neighbourhood of Belfast, where there are "branches" of the cotton manufacture, these birds use that material in the construction of their nests; and in answer to the objection that its conspicuous colour would betray the presence of the nest, and not accord with the theory that birds assimilate the outward appearance of their structures to surrounding objects, it was replied, says Mr. Thompson, that, on the contrary, the use of cotton in that locality might rather be considered as rendering the nest more difficult of detection, as the roadside hedges and neighbouring trees are always dotted with tufts of it.

The eggs are four to six in number, of a short oval form, and are usually of a pale bluish green colour. They are streaked or somewhat spotted irregularly over their whole surface with dark, dull, well-defined reddish brown spots. Some have been found of a uniform dull blue, without any spots.

Two broods are usually reared every season.

LXXXI

MOUNTAIN FINCH

BRAMBLING—BRAMBLE FINCH.

PLATE LXXXI.

Fringilla montifringilla . . . . LINNÆUS.

THE Brambling, which is a winter visitant throughout the British Isles, has only once been known to breed in a wild state in England. This occurrence was recorded by Mr. Booth, who writes: "While fishing in the west of Perthshire, in June 1866, I was forced to ascend a beech tree to release the line, which had become entangled in the branches, and while so engaged a female Brambling was disturbed from her nest, containing three eggs, which was placed close to the stem of the tree. Being anxious to procure the newly fledged young as specimens, I left her in peace; and on again visiting the spot in about ten days or a fortnight the nest was empty, and, judging by its appearance, I should be of opinion that the young birds had been dragged out by a cat."

The nest, which is usually placed in fir, birch, or other trees, about twenty feet from the ground, is formed of moss, and lined with wool and feathers or thistle-down. Mr. R. Dashwood, of Beccles, Suffolk, had these birds lay in two instances, in the year 1839, and in the latter the

3

eggs were hatched. His aviary is a large one, enclosing a considerable space of ground, and is surrounded with ivy, and planted inside with shrubs. If birds are to be kept in confinement at all, some such place is the only one in which they should be confined. The nest having been completed four days, the first egg was laid on the 16th of June in the above-named year, and another was laid each day till the 21st, when they were removed. The nest was composed of moss, wool, and dry grass, and lined with hair; and these materials were selected from a variety which the birds had the option of making use of. The foundations, which were large, were worked in among the stalks of the ivy leaves.

"In the latter part of July, in the same year," says Mr. Dashwood, writing to Mr. Hewitson, "another pair of Bramblings built, placing their nest on the ground, close to a shrub or a tuft of grass. The outside of the nest was made of moss, and it was lined with hair. From this nest I removed four eggs on the 1st of August; on the 17th of June 1840 they laid again, having built in the ivy. This nest I did not disturb, but, although the eggs were hatched, they did not succeed in rearing the young ones."

In the "Account of the Birds found in Norfolk," the authors mention the following instance, or rather instances, of these birds nesting in confinement, communicated to them by a gentleman residing near Norwich. A pair of Bramblings built a nest in an aviary in the last week in the month of June 1842, and two eggs were laid, both of which were removed and found to be good. In June 1843 the same birds again nested, and the female laid two eggs, and these having been removed, they formed a second nest in a different spot, in which four eggs were deposited.

The eggs are six or even seven in number, are rather greener than those of the Chaffinch, and like them are spotted with reddish brown.

TREE SPARROW

MOUNTAIN SPARROW.

PLATE LXXXII.

Passer montanus, ... Linnæus.

THE Tree Sparrow is irregularly distributed over England, being most common in the central and eastern counties.

Nidification commences in February, and incubation in March, two or three broods being reared in the year.

The nest is formed of hay, and is lined with wool, down, and feathers. It is loosely put together, and the consequence of this untidiness, the larger straws being left hanging carelessly outside, is that the situation of the nest is betrayed. The same situation is often again occupied from year to year.

Mr. James Dalton, of Worcester College, Oxford, informs me that he has taken the nest of this bird from a Sand Martin's hole, near Buckingham. They build in many various situations, most frequently in a hole of a tree—whence their English name—either that formed naturally by decay, or that in which some other bird, such as the Woodpecker, or one of the species, has previously domiciled; sometimes also in old nests that had been inhabited by Magpies and Crows; and in these cases, the

6

nest, that is of the Tree Sparrow, is domed over, as is done also by the House Sparrow, when it locates its habitation in similar situations. Not unfrequently they build in the thatch of barns and out-houses, but only in thoroughly country places, the entrance being from the outside; also in the tiling of houses, and in stacks and wood faggots; likewise in old walls, not many feet above the ground. Mr. Arthur Strickland, of Bridlington, has recorded that a pair built their nest, a domed one, in a hedge in the grounds of Walton Hall.

The eggs, which vary from four to six in number, are of a dull bluish or greyish white colour, speckled all over with light greyish brown of different shades. They resemble those of the House Sparrow, but are more darkly marked.

SPARROW

HOUSE SPARROW—COMMON SPARROW.

PLATE LXXXIII.

Passer domesticus,	. . .	LINNÆUS.
Fringilla domestica,	. . .	PENNANT. MONTAGU.

THE nest, which is large in size, and very loosely compacted, is usually placed under the eaves of the tiles of
houses or other buildings, in the ivy on a wall, underneath
the nest of Rook or Magpie, or in any hole or cavity that
will supply it with a convenient receptacle for its brood.
It is compiled of hay, straw, wool, moss, or twigs, and a
profusion of feathers, which the birds are sometimes seen
conveying to their holes even in winter. It often measures
as much as six inches in diameter, and sometimes even
yet more, if the situation demands it. The materials just
mentioned, as also any other that may meet the require-
ments of the bird, are variously disposed and arranged
together according to circumstances. Dove - cotes and
pigeon-houses, old walls, sheds, and ruins, are frequently
built in, and the same situation is continued to be resorted
to, and this even when the young have been exposed to
misfortune from rain ; also, as previously mentioned under
the account of the Martin, forcible possession is sometimes
taken of the nest of the latter bird. It would appear that

LXXXIII

trees are built in more from necessity than choice, namely, by yearling birds, which commence nidification late, by which time convenient places in walls have been preoccupied, or by individuals which from some cause or other had been obliged to give up the latter localities. Fewer broods in the year are produced therefore in the case of nests in trees, both from their being commenced later in the season, and from their requiring naturally more time in their construction; they are accordingly better made and larger. Mr. Meyer describes one which was handsomely built of moss, grass, and lichens, and neatly lined with hair. The entrance in these cases is by the side, and the interior is profusely lined with feathers. Three broods are often reared during the season.

The first set of eggs generally consists of four, five, or six. They are dull light grey, or greyish white, much spotted and streaked all over with ash-colour and dusky brown, varying considerably in size, shape, and colour, though preserving for the most part a general resemblance.

The lower egg on the plate, an exceptionally light brown variety, was forwarded by Mr. G. Grantham, of East Shalford, Guildford..

GREENFINCH

GREEN GROSSBEAK—GREEN LINNET.

PLATE LXXXIV.

Ligurinus chloris,	. . .	LINNÆUS.
Loxia chloris,	. . .	LATHAM.
Fringilla chloris,	. . .	TEMMINCK. SEEBOHM.

THE Greenfinch begins generally to build in April, or even earlier; the work has been known to have been completed by the 26th of March.

The nest is pretty well compacted, and much more so in some instances than in others. It is composed of small roots, twigs, moss, and straws, and lined with finer materials of the same kinds, mingled, as the case may be, with thistle-down, feathers, and hair; one was once built in the trellis-work near the drawing-room of Nafferton Vicarage, a few yards from that of the Spotted Flycatcher; but though undisturbed, it was not resorted to the following year, as was that of its near neighbour. It is placed in various situations —a low bush, or an evergreen, the ivy against a wall, or between the branches of a tree. Many nests are often found in proximity to each other in the same shrubbery; more than one sometimes even in the same bush.

The eggs, from four to six, or even seven in number, are of a pale greenish white, spotted with darker purple,

GREENFINCH.

grey, or reddish brown, and rarely streaked with brown. They do not differ much in size, shape, and colour; but sometimes the whole surface is mottled over, and again, there have been known no markings at all : the smaller end is rather pointed.

Two broods are sometimes reared in the season. The young, when fledged, fly off in a body from the nest, if approached.

The figure of the nest is from a remarkably beautiful specimen from an elm tree.

SERIN FINCH

PLATE LXXXIV.*

Serinus hortulanus,	.	.	.	KOCH.
Fringilla serinus,	.	.	.	LINNÆUS. TEMMINCK.

THE Serin breeds in Central and Southern Europe, but not, as far as is known, in the British Isles. The nest is neatly and well built of fine roots and stalks of grass, added to with spider-cots, moss, and lichens, lined with feathers and hair, or perhaps a lock or two from its "wool-stapler," the lamb, or sheep.

It is placed between the smaller branches of a shrub or small tree.

The eggs, four to five in number, are of a pale dull greenish white, with small indistinct reddish brown spots, chiefly at the larger end.

LXXXIV

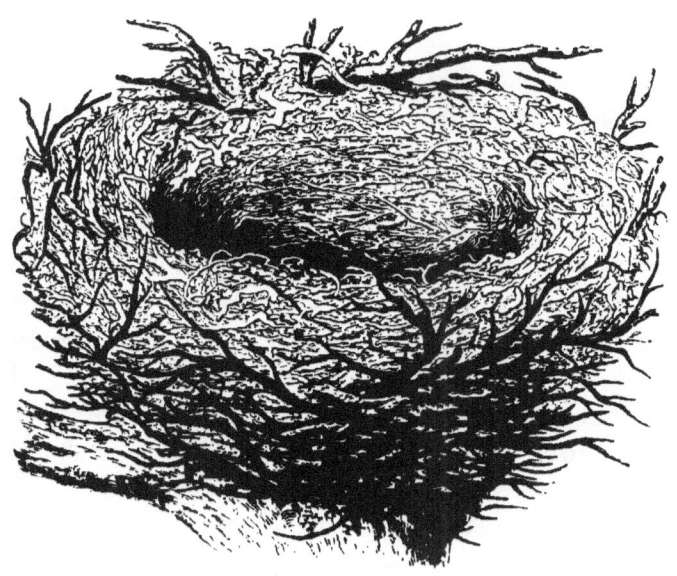

HAWFINCH

HAWFINCH

GROSSBEAK—COMMON GROSSBEAK—HAW GROSSBEAK
BLACK-THROATED GROSSBEAK.

PLATE LXXXV.

Coccothraustes vulgaris,	.	.	PALLAS. GOULD.
Loxia coccothraustes,	.	.	LINNÆUS. LATHAM.
Fringilla coccothraustes,	.	.	JENYNS. TEMMINCK.

THE nest and egg of this bird represented were taken in the parish of Beenham, Berks: the former is entirely composed of lichens and fine roots. It is frequently placed in a thorn bush, or holly tree, as also in oaks, the horse-chestnut, apple, and fir trees of the different species, at a height varying from eight or nine to thirty feet from the ground, often in a very exposed situation. It is a flattish structure built of small twigs, such as those of the oak and honeysuckle, intermixed with fragments of lichens, in greater or less abundance. The lining consists of fine roots, vegetable fibres, and a little hair, with feathers, according to Montagu. It is usually but loosely compacted.

The eggs are from four to six in number, of a pale olive or bluish green, spotted with blackish brown, and irregularly streaked with dark olive and dusky grey; some are much less marked than others, and some are of a uniform pale green.

There do not appear to be any very striking varieties. Seebohm says, "The eggs of the Hawfinch do not resemble those of any other British Finch."

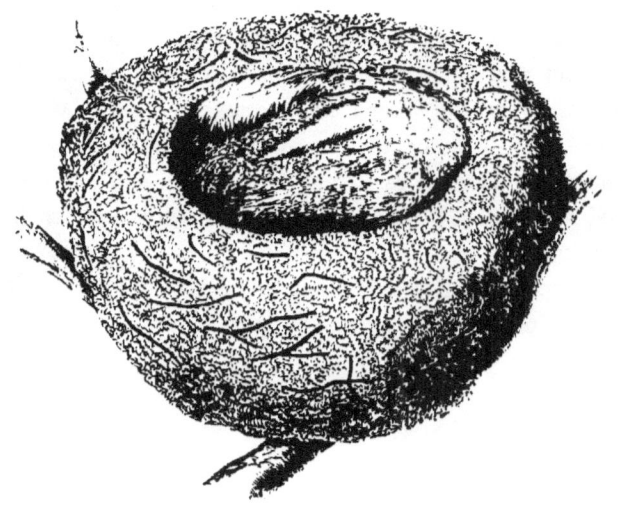

GOLDFINCH.

GOLDFINCH

PLATE LXXXVI.

Carduelis elegans, .	.	.	MACGILLIVRAY.
Fringilla carduelis,	.	.	LINNÆUS. LATHAM.

THE nest, which is a beautifully wrought structure, is placed in orchard and other trees, especially those which are evergreen, in bushes, and in some instances in hedges, and at times as much as thirty feet from the ground: it is composed externally of grass, moss, lichens, small twigs, and roots or any other appropriate substances. Inside it is elaborately interwoven with wool and hair, lined with the down of willows and various plants, and sometimes a few or more leaves or feathers. It is very neatly finished, and Bolton says is completed in three days. It is often placed in frequented situations, without much regard to passers-by. The same place is resorted to in successive years. Two broods are generally reared every season.

The eggs, four or five in number, are bluish or pale greyish white, sometimes tinged with brown, and are slightly spotted with greyish purple and brown, with occasionally a dark streak or two.

The spots vary considerably in depth of colour, from greyish purple to dark brown, or even black.

25

SISKIN

ABERDEVINE.

PLATE LXXXVII.

Fringilla spinus, ... LINNÆUS.
Carduelis spinus, ... MACGILLIVRAY.

THE nest is placed in trees, at only a short or moderate
height from the ground, about from five to eight feet or
so, or near the top of a spruce fir, and is composed of stalks
of grass, small roots and fibres, moss and lichens, lined with
hair, rabbits' fur, thistle-down, wool, or a few feathers, but
these last not as a rule. It is sometimes a furze bush within
a few feet of the ground, or in trees, when firs or birches are
usually selected.

The eggs are usually five in number, pale greenish or
bluish white, spotted around the thicker end with dull lilac
and a few reddish brown dots.

There are generally two broods in the season. Incubation
lasts fourteen days; the young are fledged and are able to
leave the nest at the end of the third week.

The Siskin has been known to build and breed in
confinement. Mr. Hewitson figures an egg which was laid,
together with three others, by a hen bird which had been
kept three years in a cage.

The Siskin nests most frequently in North Britain,

LXXXVII

though instances have also occurred in the extreme south, that is to say, in the neighbourhood of London.

Booth, a most careful observer, describes the nests as constructed in the fir trees in Scotland rather differently from other authors. He says :—" The nest has been stated to resemble that of the Goldfinch; with the exception perhaps of size, I have noticed little similarity. The outer portion is fashioned with green moss held in position by fibres of roots and strands of grass, finer materials of the same description being used for the lining, in which I have also seen a few catkins of either the birch or alder, together with a quantity of the seeds. To the best of my recollection neither wool, hair, or thistle-down, nor the flowers of the cotton-grass, were employed in any nest I examined. "

LINNET

PLATE LXXXVIII.

Linota cannabina,	.	.	.	. YARRELL.
Linaria cannabina,	.	.	.	. MACGILLIVRAY.
Fringilla cannabina,	.	.	.	. LINNÆUS. LATHAM.
Fringilla linota,	.	.	.	. LATHAM.

THE nest, of rather large size, is commonly placed in heath, grass, furze, or gorse, and is neatly constructed inside, but the outside rather roughly, being formed of small twigs, roots, straws, fibres, and stalks of grass, thistle-down, or willow catkins, intermixed with moss and wool, and lined with hair and sometimes a few feathers. It is occasionally placed in a gorse, thorn, or other bush or tree, and has been known at a height of ten or twelve feet, but is usually about four, from the ground; also in hedges; often in trees trained against the wall, particularly the pear, as affording the most concealment.

The first eggs are usually laid in April. They are from four to six in number, of a bluish white colour, spotted, most at the larger end, with purple grey and reddish brown.

18

LINNET.

Two broods are often reared in the season.

They have occasionally been known to breed in confinement, when in a good-sized aviary, and hybrids or Mules, bred between a Cock-Linnet and a female Canary, are common, and in great demand as song-birds.

REDPOLL

LESSER REDPOLE.

PLATE LXXXIX.

Fringilla rufescens,	.	.	.	VIEILLOT.
Linota rufescens,	.	.	.	NEWTON.
Linaria minor,	.	.	.	MACGILLIVRAV.
Fringilla linaria,	.	.	.	LINNÆUS. LATHAM.
Spinus linaria,	.	.	.	KOCK.
Linota linaria,	.	.	.	BUONAPARTE.
Linaria flavirostris,	.	.	.	EYTON.
Linaria rubra,	.	.	.	STEPHENS.
Linaria betularum,	.	.	.	BREHM.

TWO species of the Redpoll are usually described as distinct by British authors, but with regard to their names or their specific distinction, hardly any two writers are agreed. One of the latest authorities, Mr. Howard Saunders, speaks of the two as distinct; Mr. Seebohm regards them as mere varieties, and to them he adds the Greenland Redpoll, which has only once occurred in England. The two species generally recognised are the Lesser Redpoll and the Mealy Redpoll. The Lesser Redpoll breeds commonly in many parts of Great Britain, wherever there are woods and thickets, and rarely in any other country.

The small nest is usually built in a low bush or tree, such as an alder, hawthorn, hazel, osier, or willow, seven or eight feet from the ground, or in heather, and is fabricated

: x x x i x

of moss, hair, wool, and stems of grass, lined with willow catkins, or feathers. I have been informed by Mr. F. Wise, of Arram, in Holderness, of one at a height of about twenty feet, on the end of a bough of an oak, among a tuft of small shoots where the end had been broken off. It is rather carefully and neatly constructed. Several nests are often built quite near together.

This species lays from four to six eggs: their colour is pale bluish green, spotted with orange brown, principally towards the larger end, and sometimes a few thin streaks of a darker colour—brown or black.

The nest is figured from a specimen which was taken on the 6th of June, in the year 1853, in the neighbourhood of Driffield. It is made of the usual materials, as mentioned in the description. It contained three eggs, from one of which the engraving was drawn.

The Redpoll is a late breeder, eggs seldom being taken before the beginning of June.

MEALY REDPOLL

PLATE XC.—FIGURE I.

Linaria canescens,	GOULD.
Linaria borealis,	SELBY.
Linaria minor,	SELBY.
Fringilla linaria,	LINNÆUS.

THE Mealy Redpoll is an irregular winter visitant to the east coast, breeding only in the birch woods of the north, and straying to the south only in the winter. In confinement it has bred with the last species, but has not been known to nest in a wild state in England.

The egg is described by Meyer as being pale greenish blue, sprinkled over with pale but distinct spots of a reddish brown colour, some of them inclining to lilac, chiefly confined to a zone around the larger end.

TWITE

MOUNTAIN LINNET.

PLATE XC.—FIGURE II.

Fringilla flavirostris,	. . . . .	LINNÆUS.
Linaria montana, .	. . . . .	SELBY.

THE nest of the Twite is generally built in heather or on the ground in grass, growing corn, or by the side of a furrow, sometimes in low bushes. It is formed of small roots, twigs, and stalks of shrubs, heather, moss, and dry grass, and is lined with a small quantity of hair or wool, and a few feathers. A pair built and reared their young in the aviary of Mr. Thomas Walker, of Rosebank, near Tunbridge Wells.

The eggs, four, five, or occasionally six in number, are of a pale greenish or bluish white, spotted with reddish brown, or light brown and purple-red towards the larger end, with sometimes a few blackish dots, and are scarcely to be distinguished from those of the common Linnet.

BULLFINCH

PLATE XCI.

Loxia pyrrhula,	.	.	.	LINNÆUS.
Pyrrhula vulgaris,	.	.	.	FLEMING. SELBY.

TOWARDS the end of April the birds pair, and nidification is commenced in the beginning of May, and is finished by the end of that month, or the early part of June.

The nest, small and shallow, is formed of small twigs, commonly those of the birch, beech or hornbeam, and is lined with fine roots, grass, wool, and hair, the whole being rather carelessly built and not firmly compacted; in some instances moss is added. The middle part is more carefully finished and of the finest of the materials. It is generally placed low, either in a tree, such as a fir, or in the middle of a bush or other underwood, frequently a hawthorn, at a height of four or five feet from the ground. It is often built in an evergreen, a cover, plantation, or shrubbery, even near a house, commonly in a wood, and occasionally, though but seldom, in a garden. They will breed in confinement, especially if they have ample room.

The eggs, four to six in number, are clear greenish blue, speckled and streaked with purple grey, and dark

24

BULLFINCH

purple. They are hatched towards the end of May, after an incubation of fifteen days. The male takes his turn in sitting with the female. The latter sits very closely, though she is in general easily frightened away. The male is less so, but it is said that if he be disturbed the nest is almost always deserted, which is not the case when the female is alarmed. Mr. W. Read, of Hayton, has recorded that, when resident at Frickley Hall, near Doncaster, a hen bird which built in a laurel near the house suffered herself to be touched while sitting on her young ones, and would feed from the hand without the least fear. The birds are supposed to pair for life; the members of the family keep together until the spring. Two broods are frequently reared in the season.

SCARLET BULLFINCH

SCARLET GROSSBEAK.

PLATE XCII.—FIGURE I.

Pyrrhula erythryna,   .   .   .   PALLAS.

THIS species, which is a rare wanderer to the west, breeds in Finland and Russia.

The nest is loosely built, but neatly lined inside with the finer stems of plants, interwoven with a few hairs.

The eggs are from four to six in number; their colour a deep greenish blue, with a few dark reddish-brown or nearly black spots and specks, with blots of pale purple red.

XCII

PINE GROSSBEAK

PINE BULLFINCH.

PLATE XCII.—FIGURE II.

Loxia enucleator,	. . . .	LINNÆUS.
Pyrrhula enucleator,	. . . .	SELBY. JENYNS.

THE Pine Grossbeak is a rare visitant, never nesting in Great Britain.

The nest is made of small birch sticks, and is lined with fine stiff grass or lichen. It is usually placed on the branch of a fir or birch tree, only a few feet above the ground.

The eggs are three to four in number, deep greenish blue in colour, spotted with brownish purple.

CROSSBILL

COMMON CROSSBILL—SHELL-APPLE.

PLATE XCII.—FIGURE III.

Loxia curvirostra, . . . LINNÆUS. LATHAM.

THE Common Crossbill is generally seen in flocks in England from autumn to spring, but occasionally remains to breed, though usually nesting in the pine forests of the north of Europe.

Nidification commences very early, even in February or March. According to Temminck second broods follow the earliest. Several often build together.

The nest is placed in the angle of the junction of the branches of a tree, naturally the fir, but the apple also has been known, and is loosely compacted of small twigs, grass, straws, stalks, and moss or lichens, according with the colour of the tree it is placed on, lined with softer materials, such as hair, moss, wool, or feathers. They have been known only about five or ten feet from the ground, and up to forty. The edges of the nest extend from three to five inches beyond the middle part.

The eggs, four or rarely five in number, are white or greyish white, spotted, chiefly at the thicker end, with red-brown, reddish, bluish red, purple, or brown; in some well-defined, in others shaded off, and in others only lines.

PARROT CROSSBILL

PLATE XCIII.

Loxia pityopsittacus, . . . Bewick. Fleming.

THE Parrot Crossbill is now regarded by all recent writers as merely a large stout-billed race of the last species. These larger birds are seldom seen in the British Islands.

The nest is placed chiefly in lofty forest trees, and is composed of small twigs, lined with dry grass or leaves of the fir tree.

The eggs are said to be four or five in number, ash-coloured, or bluish white, and spotted with bluish red and dusky at the larger end.

The Rev. H. B. Tristram has obligingly forwarded to me the egg made use of for the plate. It was obtained by himself at Hennsand, in Sweden.

WHITE-WINGED CROSSBILL

PLATE XCIV.

Loxia leucoptera, . . . GMELIN. BUONAPARTE.
Loxia falcirostra, . . . PENNANT. FLEMING.

THE White-winged Crossbill of North America has on a few occasions occurred in Great Britain. It is generally regarded by recent writers as identical with the Two-barred Crossbill of the north of Europe (*Loxia bifasciata*), small flocks of which have occasionally occurred in the eastern counties.

Very little is known of the incubation of this bird, but the nest has been described as composed of lichens, spruce twigs, coarse hairs, and shreds of fine bark; it is placed on the branches of pine trees.

The eggs, five in number, are described as white, marked with yellowish spots; otherwise, as pale blue with fine dots of black and lilac grey.

30

XCV

ROSE-COLOURED PASTOR

ROSE OUZEL—ROSE-COLOURED OUZEL—ROSE-COLOURED
STARLING.

PLATE XCV.

Pastor roseus, ... FLEMING. SELBY.

A N irregular visitant to this country, breeding in Western Asia.

The nest is located in the holes of trees, and in
cavities in old walls or rocks, as also on the ground,
numbers of nests being built together. A few straws, sticks,
wool, leaves, grasses, roots, mosses, feathers, and such like
are the materials, if any, for often the eggs are laid on the
bare ground, or in a deep hollow or hole under rocks
and stones.

The eggs are five or six in number, are glossy bluish
white, with the like faint tinge of green blue or blue
green.

STARLING

STARE—COMMON STARLING.

PLATE XCVI.—FIGURE 1.

Sturnus vulgaris, ... LINNÆUS.

NIDIFICATION commences about the end of March or the beginning or middle of April. They build in church-steeples and in hollows and eaves of the walls of houses, castles, spires, towers, or ruins, as also in those of trees, as well as in cliffs and rocky and precipitous places; at times in dove-cotes and pigeon-houses, in caverns and under rocks, and even have been known to occupy the holes deserted by rats, more or less fashioned for themselves. Where any or all of these are wanting, the abutment of a bridge, or any suitably high building, is utilised. A rabbit-burrow is also sometimes resorted to, or the hole in a tree scooped out by a Woodpecker. In Woburn Park, Bedfordshire, I am informed by Mr. George B. Clark that Starlings have built some dome-shaped nests in Scotch firs, the entrance placed near the branch of the tree, the nests being made of coarse grass, and lined with fine grass. He also writes that he has known them occupy holes previously excavated by Sand Martins for themselves. Mr. J. M'Intosh, describing a famous chestnut tree in the grounds of Canford House, Dorsetshire, one of five planted by John

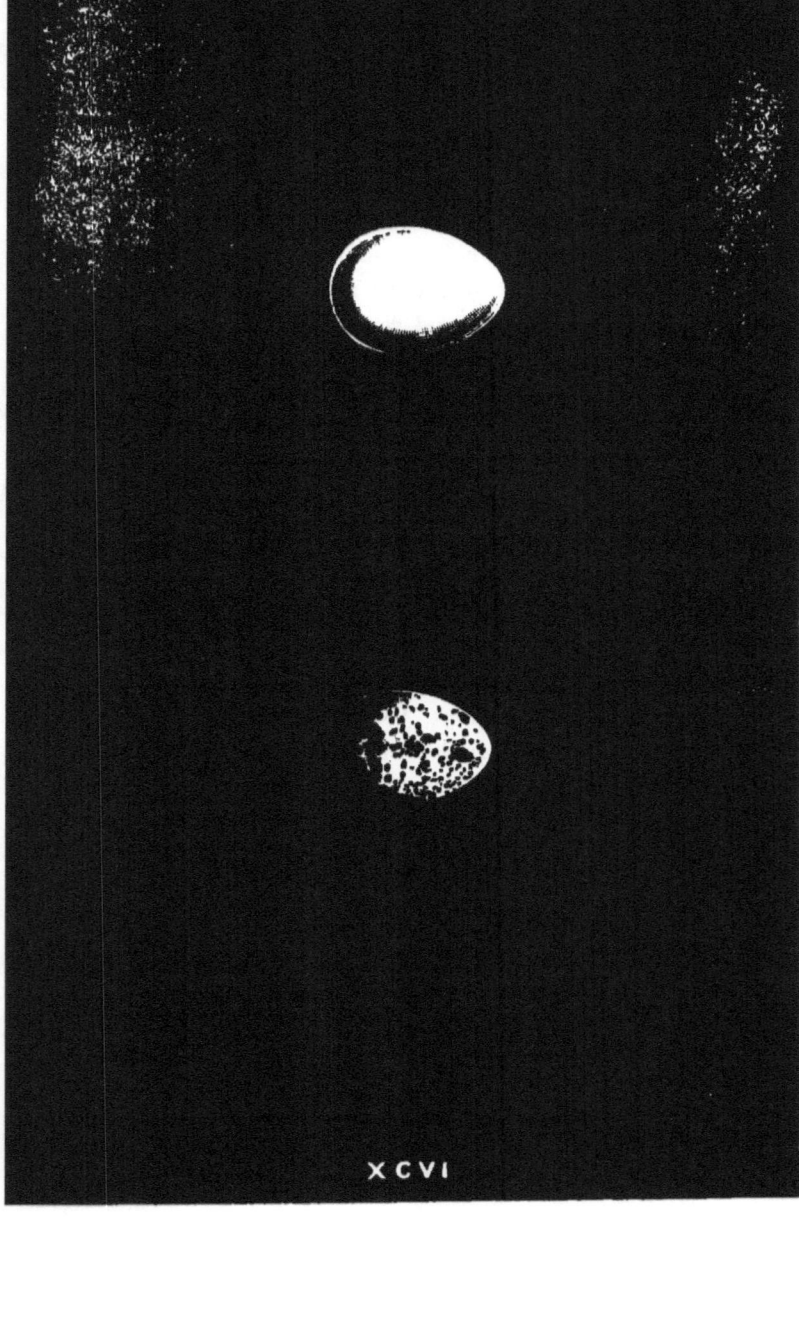

XCVI

of Gaunt, and still flourishing so long since "time-honoured Lancaster" himself has mouldered into dust, that at its base was a colony of rabbits, in the trunk a nest of cats, and immediately above the latter, one of Starlings. One has been built for two or three years in succession, in the garden of Nunburnholme Rectory, in the depth of a hole in a large old acacia tree.

The nest is large, and fabricated of straw, roots, portions of plants, and dry grass, or hay, with a rude lining of feathers and hair. The birds will sometimes resort most pertinaciously to the same building-place. In one instance the eggs are said to have been found in the nest of a Magpie. One pair having with much difficulty forced their way into a ball used by being raised or let down to act as a signal on a railway, there built their nest, and though the ball was elevated and lowered to within a few feet of the ground fourteen times a day, this did not interfere with their proceedings, and in due time four eggs were laid with every prospect of being duly hatched. This was near Kilwinning, on the Ardrossan line, in 1853, and the circumstance was recorded in the *Dumfries Courier*. The hen sits very close, is fed by her mate, and has been known to allow herself to be taken by the hand from the nest.

The eggs, four to seven in number, are of a delicate pale blue or blue green colour: some have a few black dots. Mr. R. J. Davidson, of Muirhouse, informs me of a nest of five white eggs, which he found in a hollow tree at Dedham, in Essex, in 1862. Mr. G. Warren, of Witnesham Vicarage, near Abingdon, found a nest with the eggs all but pure white, and forwarded me two of them as specimens.

Incubation lasts about sixteen days; both birds feed the young. Two or three broods may be raised in the year in some places, but ordinarily one, or at the most two.

Mr. J. R. Fisher states that Mr. Gurney told him of a Starling, the young of which having been taken from the nest and placed in a cage which was hung upon a wall, were discovered and fed by the old bird until they were able to fly, at which time, and not before, she unfastened the door of the cage and let them out.

RED-WINGED STARLING

RED-WINGED BLACKBIRD — RED-WINGED MAIZE-BIRD—
MARSH BLACKBIRD—SWAMP BLACKBIRD.

PLATE XCVI.—FIGURE II.

Agelæus phæniceus, .	.	NEWTON.
Sturnus prædatorius,	.	LUBBOCK. WILSON.
Icterus phænicurus,	.	BUONAPARTE.

THIS bird has been found about a dozen times in the British Isles. It is common throughout the United States. The English specimens have most probably escaped from confinement.

About the middle of April the birds pair, and nidification commences the last week in April, or the beginning of May, or even later, according to the latitude in which they happen to be.

The nest is placed variously in a bush or tree, a few feet from the ground, or in a tussock of rushes or tuft of grass, or even, and not unfrequently, on the ground. It is composed of rushes and long tough grass, and lined with finer portions of the latter; the rushes are interlaced among the surrounding twigs, if in a tree, or among the rushes, if on the ground, in which latter case the whole structure is less elaborate than in the former. Several nests are often built in the immediate neighbourhood of each other.

35

The eggs, about five in number, are of a pale bluish-white colour, encircled at the larger end with spots and streaks of dark reddish brown, with a few others scattered here and there, and some faint blots of purple grey and lines and dashes of black.

DIPPER.

DIPPER

Cinclus aquaticus,　　　　　　　Fleming. Selby.

NIDIFICATION begins about the middle of April. The nest, which is cleverly concealed and large, measuring ten or twelve inches in diameter and seven or eight in depth, being domed, is well compacted of moss and grass, and fully lined with leaves. It is placed in some cavity in a rock, or under the protection of some overhanging stone in the immediate neighbourhood of the rippling stream or murmuring waterfall, the birds' favourite haunt. Different specimens, however, vary in size as well as shape, adapted doubtless to the circumstances of the spot they are placed in, some being a couple of inches less than the size just spoken of. The aperture is in front, from three to four inches in width, and about one and a half in height. Macgillivray mentions one, described by Mr. Thomas Durham Weir, which was built in an angle between two fragments of rocks under a small cascade, and although the water fell upon part of the dome, the compactness with which it was put together rendered it waterproof.

Another was similarly placed in a hole of a wall close to a waterfall, which passed over it, but the birds nevertheless

37

flew in and out with the greatest ease. Again, another placed
for several years in succession on the rafter of a salmon fish-
lock is recorded by Mr. Hewitson. Others have been known
to be built within the passage of a mill-race. In such cases
the mother bird will often dash two or three times through
the rushing waters, as if in the enjoyment of pastime, before
resuming her place on the eggs. The young soon quit the
nest, and are at home almost from the first in the water.

The birds are strongly attached to their accustomed
building-place, and the same spot has been known to be
occupied for thirty-one years.

The dipper rears two or three broods in the year. A
second clutch of eggs is often deposited in the same nest
with the young birds.

The eggs, from four to six in number, are white, and of
a regular oval form.

MISSEL THRUSH.

MISSEL THRUSH

MISSELTOE THRUSH—STORM-COCK—MISSEL-BIRD—SHRITE—
SHRIKE-COCK—HOLM THRUSH.

PLATE XCVIII.

Turdus viscivorus, . LINNÆUS.

THE Missel Thrush nests very early, often commencing even in February, and nests with eggs have been found early in March.

The nest, which is a loose structure, is a compilation of twigs, small sticks, straws, grasses, leaves, lichens, wool, or mosses, compacted inwardly with mud, intermingled with still smaller roots, finer grasses, and moss, or indeed any soft material, feathers in some cases, frequently with grass alone; sometimes the outside is partly covered with lichens and mosses, the former taken from or resembling those on the tree itself, to which they consequently give the fabric verisimilitude. The width is about four inches and a half, the depth two and three fourths, and the thickness of the sides an inch and three quarters. Mr. Hewitson mentions a nest of which the foundation was of mud, strongly cemented to and nearly encirling the branches between which it was fixed. This material appears to be occasionally used a little with the lining. It is often placed in very exposed situations in the hollow caused by the divergence of the branches from the trunk, at a height of ten or fifteen feet from the ground. Shy, too, as the bird is at other times, in its nidification it is not deterred from any appropriate situation by the near propinquity of a house, or by persons constantly passing and repassing, it often building without any attempt at concealment. The same tree is often returned to year after year, if the birds be undisturbed, and Mr. Frederick Bond, of Kingsbury, has known the same nest used twice in the same season. They will suffer other species to build near them, without any molestation, even during the time of incubation.

The eggs are four to five in number, of a greenish or tawny white colour, spotted and blotted and more or less suffused irregularly with brown, reddish brown, or purple red, they vary in size as well as in colour and shape, some being much larger than others. They are hatched in about sixteen days.

Two and sometimes three broods are reared in the season.

XCIX

FIELDFARE

FELDFARE—FELT—FELTFARE—BLUE-BACK—BLUE-FELT.

PLATE XCIX.

Turdus pilaris,   .   .   LINNÆUS. LATHAM.

FIELDFARES, which are only winter visitants to Britain, breed in forest regions of Northern Europe in large colonies, as many as two hundred nests and upwards having been found within a small circuit of the forest. The same situations appear to be resorted to from year to year, as with the Rooks.

The nest, which is placed in pine or fir trees, at a height of from four to forty feet from the ground, is made of small sticks, grass, and weeds, cemented together with a small quantity of clay, and lined with fine grass. It is for the most part placed on the branches of the birch, alder, or pine, against the trunk of the tree, but sometimes at a considerable distance from it, towards the smaller end of the thicker branches. Single nests, however, sometimes occur.

The eggs are from four to five or six in number, of a pale bluish green, of different shades, spotted, mottled, and streaked with darker or lighter reddish brown. "They are sometimes so closely freckled over that the colour of the freckles predominates; and there is a variety in which the ground colour is most seen, the red-brown spots being larger and much more sparingly sprinkled." They vary considerably, as may be observed from the two types represented.

Two broods are generally reared each season.

Unfinished nests have been found, and others with newly-laid eggs in them, so late as the 30th of May.

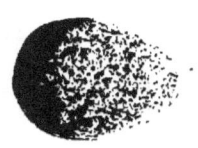

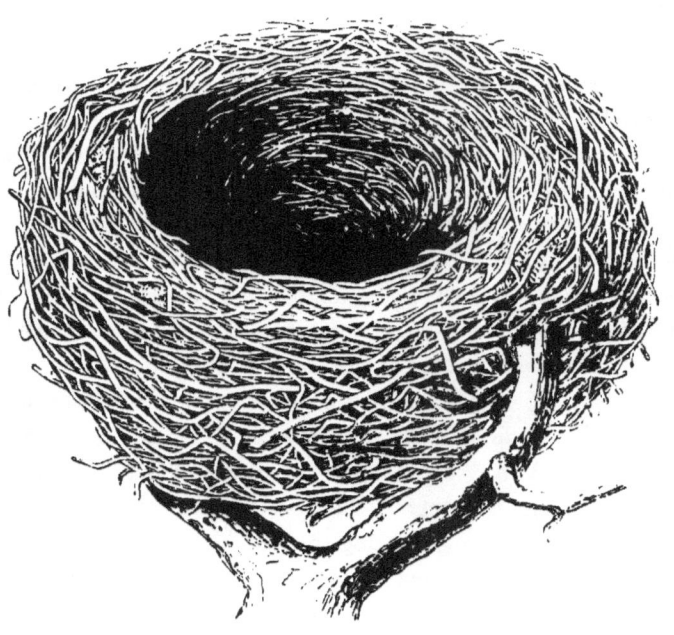

REDWING.

REDWING

SWINEPIPE—WIND THRUSH.

PLATE C.

Turdus iliacus, . Linnæus.

THE Redwing is another winter visitor, nesting only in the pine forests of the Arctic Circle. The nest is placed in the centre of a thorn or other bush, alder, birch, maple, or other tree, or a cluster of stems, and is made of moss, roots, and dry grass outwardly, cemented together with clay, and lined inwardly with finer grass.

Mr. Wolley says that this bird "makes its nest near the ground, in an open part of the wood, generally in the outskirts, on a stump, a log, or the roots of a fallen tree; sometimes amongst a cluster of young stems of the birch, usually quite exposed, so as almost to seem as if placed so purposely—the walls often supported only by their foundation. The first or coarse part of the nest is made for the most part of dried bents, sometimes with fine twigs and moss; this is lined with a thin layer of mud, and then is added a thick bed of fine grass of the previous year, compactly woven together, which completes the structure. Outside is often a good deal of the kind of lichen called reindeer moss, and one nest particularly, which I have preserved, is entirely covered with it; when it was fresh, and the fine ramifications of the lichens unbroken, it had a most beautiful appearance."

The eggs are said to be found at the end of May, or June, and to be towards six in number; they are of a pale bluish green, spotted or streaked with reddish brown.

Two broods are frequently reared during the season.

THRUSH.

THRUSH

THROSTLE—SONG THRUSH—COMMON THRUSH—MAVIS.

PLATE CI.

Turdus musicus,	.	LINNÆUS.
Merula musica,	.	SELBY.

THE Thrush usually commences to build in the latter end of March, and the eggs are deposited earlier or later in April, though sometimes not until May, according to the season.   Nests are known to have been begun even so early as the middle of February, and young birds have been found in March.   Mrs. Harriet Murchison, of Bicester, Oxfordshire, has forwarded me a specimen of a nest with four eggs, which was found at that place on the 6th of January 1853.   A second brood is generally reared in the season.   The female is extremely attentive to her charge, and will sit on the nest until quite closely approached, and will sometimes suffer herself to be taken sooner than forsake it.   If disturbed or alarmed, she will testify her anxiety by flying round with ruffled feathers and outspread tail, uttering a note of alarm, and violently snapping the bill.

The nest is a very bulky structure, composed of moss, small twigs, straws, leaves, roots, stems of plants, and grass, lined with a thick layer of clay and decayed wood.   It is placed in a hedge or thick bush of any kind at a small height from the ground, and likewise at times on a rough bank among moss, brambles, or shrubs, as also, where the country is unwooded, on the level ground, at the most under the shelter of some projecting stone or crag, in the crevice of a rock, or in a tuft of heath.

Mr. John H. Blundell, of Luton, Bedfordshire, informs me that he has found the nest of a Thrush in the side of a wheat stack. The Rev. W. Waldo Cooper, of West Rasen, Lincolnshire, records in the *Zoologist*, page 1775, that he has found one on the ground, three feet from the nearest bush; and Mr. John Barlow relates a similar instance.

The eggs, usually four or five in number, are of a beautiful clear greenish blue colour, with distinct black or rusty brown spots and dots, principally over the larger end. Unspotted varieties are not very uncommon. The eggs vary considerably in size, some being very small.

WHITE'S THRUSH

PLATE CII.—FIGURE I.

Turdus varius .   PALLAS.

WHITE'S Thrush is a rare accidental visitor to Europe, breeding in North-eastern Siberia, and wintering in Japan.

The nest is composed, according to Swinhoe, of withered rushes, grass, and moss, lined with mud and rootlets.

The eggs are greenish white, with minute reddish spots.

ROCK THRUSH

PLATE CII.—FIGURE II.

Turdus saxatilis,	.	.	LINNÆUS.
Petrocincla saxatilis,	.	.	VIGORS. GOULD.

THE Rock Thrush is a rare visitor, breeding in the south of Europe. The nest is made of moss, roots, and dried grass, without clay. It is placed in crevices of rocks, walls, or ruins, occasionally in a tree-stump.

The eggs are described as being four to five in number, and of a pale greenish-blue colour, sometimes slightly speckled with light brown.

Two broods are often reared in the year.

BLACKBIRD

PLATE CIII.

Turdus merula, ...	Linnæus.
Merula vulgaris, ...	Selby. Gould.

THIS species pairs in February or March, but occasionally much earlier. A nest with eggs has been found
in January.

The nest is placed in a variety of situations, and is
frequently found in a heap of sticks, even though placed in
an outhouse, or most commonly in a bush; sometimes in a
tree against a wall, or in a tree or wall covered with ivy;
an instance has been known of its being placed on the stump
of a tree, close to the ground, and Sir William Jardine
found one on the ground, at the foot of a tree; another was
also seen in a similar situation, at the foot of a hazel bush,
in a wood, by the Rev. W. Waldo Cooper, of West Rasen,
Lincolnshire: in the same wood he saw another on the
stump of a hazel which had been cut down, and from which
several stems had grown; it was not raised an inch from the
ground, but was quite surrounded by the new branches; and
others on the ground have been recorded. It is often put
in a hedge, at a height of three or four feet; and sometimes in a hole in a wall or rock. It is made of roots,

small twigs, and stalks of grass, with perhaps some lichens or fern, and is covered on the inside with mud, and lined with finer parts of the other materials and grass; it is sometimes most admirably hidden, so as almost to baffle detection. It is at times placed on the top of a fence or the summit of a wall. The same situation is occasionally resorted to from year to year. The female sits for thirteen days.

The eggs are commonly five in number, sometimes four, and sometimes, though but rarely, six; they are of a dull light greenish blue, mottled and spotted with pale reddish brown, the markings being closer at the larger end, where they sometimes form an obscure ring. They vary very much in size as well as in shape and colour. Mr. Hewitson, in his "Coloured Illustrations of the Eggs of British Birds," figures one elegantly covered over at the larger end with minute reddish-brown specks, and likewise, but less thickly, over the remainder — the green showing through; and a second curiously marbled with irregular dashes and specks of reddish brown over the green colour. Another variety is similar to the last, except that the ground colour is lighter, and the spots smaller. Another, in his possession, clear spotless light blue, with the whole of the larger end suffused with reddish brown. Mr. J. B. Ellman, of Battel, relates in the *Zoologist*, page 2180, that he had an egg in which the spots were at the smaller end. Some of the eggs are much larger than others, and they also vary much in colour and markings, as well as in shape, some being much more round, and others much more oval, than others: in some instances, the smaller end is rounded and obtuse. Two and sometimes three broods are reared in the season.

Booth in his "Rough Notes" relates an interesting account of two Blackbirds constructing no less than five nests in the season, in a thick bush of cypress. He says :— "In a garden near Brighton I noticed, in 1880, two Blackbirds (they could scarcely be termed a pair) construct no less than five nests during the season. In every instance the nests were placed in a thick bush of *Cupressus macrocarpa*. The first brood when about a week old, early in March, were dragged out and killed by a cat. On Saturday, May 1st, the second brood died in the nest through exposure to the cold east winds, and on the following Monday the third nest was commenced. On the 12th the old male was unfortunately caught in a cat-trap (set for their especial preservation), and so badly nipped that death must have been instantaneous. The female, however, was not inconsolable, and within a day or two, without the slightest intermission to her family arrangements, a new mate was found. All went smoothly for the future, and three broods were now successfully reared."

RING OUZEL

ROCK OUZEL—RING THRUSH.

PLATE CIV.

Turdus torquatus,	. . . .	LINNÆUS.
Merula torquata,	. . . .	SELBY. GOULD.

THE nest of the Ring Ouzel is usually built among the heather or ling in a hollow on the ground. It is hidden more or less by a tuft of heath, the root of a tree, a large stone, or a projection of the rock on which it is placed : those found in the more southerly counties were placed at a height of about five or six feet from the ground in such a situation as a small bush or stunted tree. It measures about seven inches in diameter, about three and a half in depth on the outside, and about two inches inside. It is composed of dry grasses, heather, stems, or stalks, thickly matted together, with here and there an occasional leaf; on the inside it is lined with mud, within which is another lining of fine grass. When the young are hatched, the parent birds naturally fly at and about any intruder.

The eggs are pale greenish blue, sparingly freckled with pale purple and reddish-brown markings. They are four or five in number. A second brood is frequently hatched in July.

5

GOLDEN ORIOLE.

GOLDEN ORIOLE

PLATE CV.

Oriolus galbula, ... Linnæus

THE Oriole, though but a straggler, has occasionally nested in the south of England, and would doubtless do so habitually if not so constantly persecuted.

The nest is usually suspended from the small forked bough of a tall tree, to which it is firmly attached. It is made of stalks of grass, small roots, and wool, cleverly interwoven together, and is lined with the finer portions of the same materials.

The eggs are commonly four or five in number, of a clear white colour, blotched with reddish purple.

ALPINE WARBLER

ALPINE ACCENTOR—COLLARED STARE.

PLATE CVI.

Accentor alpinus,	.	.	.	.	GMELIN.
Accentor collaris,	.	.	.	.	NEWTON.
Sturnus moritanicus,	.	.	.	GMELIN.	LATHAM.

A RARE straggler to England, breeding in Central and Southern Europe.

The nest is placed on the ground among stones, or in some cavity or crevice of mountain rock, as also under the shelter of the alpine rose or other low bush. It is made of fine grass, roots and lichens, and is lined with moss, wool, and hair.

The eggs, four or five in number, are of a beautiful light greenish-blue colour; they are unspotted. There are said to be two broods in the year.

ALPINE WARBLER.

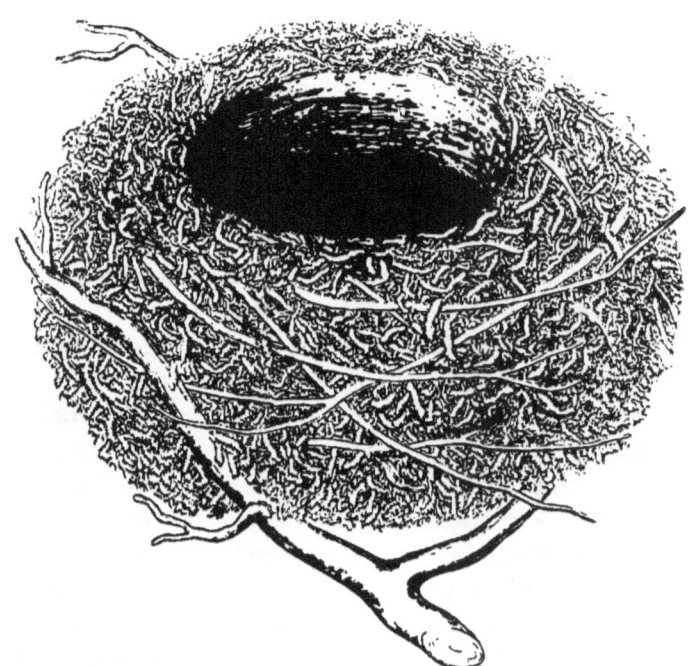

DUNNOCK.

DUNNOCK

HEDGE-SPARROW—SHUFFLE-WING—HEDGE-WARBLER—
HEDGE-CHANTER.

PLATE CVII.

Accentor modularis, ... Linnæus.

THE nest of the Hedge-Sparrow is generally placed in
hedges, low furze or other bushes, or shrubs, a few
feet from the ground, but also, in lack of these, in holes
of walls, stacks of wood, in the ivy against a wall, and
other similar places. The Rev. Charles Forge, of Driffield,
records in the *Zoologist* that he found one among the
small branches of an elm tree, standing apart from any
hedge. It was placed close to the bole or trunk of the
tree, at about ten feet from the ground. Exteriorly, it was
composed of wheat straw, intermingled with small recently
dead twigs of the elm, to which the dried leaves were still
attached. It had no other lining than the green moss commonly used by the Hedge-Sparrow in the construction of its
nest, and contained a single egg. A pair built and reared
their young in the aviary of Mr. Thomas Walker, of
Rosebank, Tunbridge Wells.

The nest is deep and well rounded, from four and a half
to five inches in diameter on the outside, and nearly two
inches deep. It is made of small twigs, roots, and grass,

55

lined with moss, and then with hair, grass, wool, or down, or any appropriate substances at hand.

The eggs, which are often seen so early as the beginning of April, are four or five, rarely six in number, and of a very elegant greenish-blue colour. Mr. Archibald Hepburn, records in the *Zoologist* his having seen an egg of this species, which was thrown out of the nest by the parents, and was of a bluish white colour, mottled and speckled with light brown; it was much rounder than the usual shape, and was empty inside.

Two broods are reared in the year; preparations for one being made about the middle of March, and for the latter, at the beginning of May : three are sometimes hatched.

Meyer, in his "British Birds," mentions his having seen a nest on the 21st of January, and that he found one with a newly-laid egg in it on the 22nd of July. The same situation is frequently resorted to from year to year.

The Rev. H. A. Macpherson records, in his "Fauna of Lakeland," that in 1888 he found a brood which did not leave the nest until the 8th of September..

REDBREAST.

REDBREAST

ROBIN—ROBIN REDBREAST—RUDDOCK—ROBINET.

PLATE CVIII.

Erythacus rubecula,	.	.	NEWTON.
Sylvia rubecula,	.	.	NAUMANN.
Motacilla rubecula,	.	.	MONTAGU. BEWICK.

THE Robin, as the Redbreast is familiarly termed, nests very early in the spring, and the eggs are usually laid about the beginning of April; but young birds have often been found in the nest by the end of March. In backward seasons they are usually later. Macgillivray, writing in Scotland, mentions one seen on the 9th of May 1831, and another on the 2nd of June 1837, which he believed to be the first brood of that year. A Robin's nest, containing several eggs, was taken near York the first week in February 1844, there being snow on the ground at the time; another, which had five eggs, was found at Moreton-in-the-Marsh in the second week of January 1848; another, with the like number of eggs, in a garden at Wheldrake, near York, the 10th of the same month; and one, also with eggs, near Belfast on the 20th of February 1846. A nest with two eggs, on which the hen bird was sitting, was found near the end of November 1851, at Gribton, Dumfriesshire, the seat of Mr. Francis Maxwell.

The nest of the Robin, which is built of fine stalks, moss, dried leaves, and grass, and lined with hair and wool, with sometimes a few feathers, is generally placed on a bank under the shelter of a bush, or sometimes in a bush itself, at a low height from the ground, and occasionally in a hole in a wall covered with ivy, a crevice in a rock, among fern and tangled roots—the entrance perhaps being through some very narrow aperture, or an ivy-clad tree. It measures about five inches and three quarters across, and two and a half in internal diameter. It is concealed with great care and success.

King William the Fourth had a part of the mizzen-mast of the *Victory*, against which Lord Nelson was standing when he was mortally wounded, placed in a building in the grounds of Bushey Park when he resided there. A large shot had passed through this part of the mast, and in the hole it had left, a pair of Robins built their nest and reared their young. The relic was afterwards removed to the dining-room of the house, and is now in the armoury of Windsor Castle.

A loft is frequently built in, and in one instance, the nest having been obliged to be removed for an alteration in the wall, the hen bird did not forsake it, though placed elsewhere. Another nest was placed on a shelf in a pantry, among some four-sided bottles, so that it was made of a square shape. When disturbed by the entrance of any person, the bird alighted on the floor till the visitor had gone, when it immediately returned to its nest.

Mr. Jesse relates the following :—"A gentleman had directed a waggon to be packed, intending to send it to Worthing, where he himself was going. For some reason his journey was delayed, and he therefore directed that the waggon should be placed in a shed in the yard, packed as it was, till it should be convenient for him to send it off. While it was in the shed, a pair of Robins built their nest among some straw in it, and had hatched their young just before it was sent off. One of the old birds, instead of being frightened away by the motion of the waggon, only left the nest from time to time for the purpose of flying to the nearest hedge for food for its young, and thus alternately affording warmth and nourishment to them, it arrived at Worthing. The affection of this bird having been observed by the waggoner, he took care in unloading not to disturb the Robin's nest; and the Robin and its young returned in safety to Walton Heath, being the place from whence they had set out; the distance travelled not being less than one hundred miles. Whether it was the male or female Robin which kept with the waggon I have not been able to ascertain; but most probably the latter; for what will not a mother's love and a mother's tenderness induce her to do?"

The eggs, generally five or six in number, are usually of a delicate pale reddish white, faintly freckled with rather darker red, most so at the larger end, where a zone or belt is sometimes formed. Some are entirely white, without a trace of marking, whilst others are so clouded with spots as to hide the ground colour, and some deeply blotched and streaked with dark reddish brown.

I may here perhaps make the following quotation from my "History of British Birds:"—

"Gentle reader, if indeed you be of gentle blood, and will read the following touching lines of the poet Thomson, descriptive of the return of a bereaved parent bird to her robbed home, if ever you have plundered a Robin's nest

or that of any other bird, let me hope that you will 'steal
no more : '—

> ——'To the ground the vain provision falls!
> Her pinions ruffle, and, low drooping, scarce
> Can bear the mourner to the poplar shade,
> Where, all abandoned to despair, she sings
> Her sorrows through the night, and on the bough
> Sole sitting, still at every dying fall
> Takes up again her lamentable strain
> Of winding woe, till wide around the woods
> Sigh to her song, and with her wail resound.'

" Here is no 'poetic license,' but if you think there is,
the following well-written 'plain prose' of the amiable Mr.
Jesse will satisfy the possible doubt : — 'I had an opportunity,' he writes in his ' Gleanings in Natural History,'
'this summer of witnessing the distress of a Robin when,
on returning to her nest with food for her young, she
discovered that they had disappeared.   Her low and
plaintive wailings were incessant.   She appeared to seek
for them among the neighbouring bushes, now and then
changing her mournful cry into one which seemed like a
call to her brood to come to her.   She kept the food in
her mouth for a short time, but when she found that her
cries were unanswerable, let it fall to the ground.'

" So also Virgil, though speaking of a different species,
in his Fourth Georgic—for Nature was the same eighteen
hundred years ago as she is now—

> ' Qualis populeâ mærens philomela sub umbrâ
> Amissos queritur fœtus, quos duros arator
> Observans nido implumes detraxit : at illa
> Flet noctem, ramoque sedens, miserabile carmen
> Integrat, et mæstis late loca quæstibus implet.'

" Thus well rendered by Dryden—

' So, close in poplar shades, her children gone,
The mother nightingale laments alone,
Whose nest some prying churl had found, and thence
By stealth conveyed the unfeathered innocents :
But she supplies the night with mournful strains,
And melancholy music fills the plains.' "

BLUEBREAST

PLATE CIX.

Cyanecula suecica	LINNÆUS.
Sylvia suecica	JENYNS.
Phœnicura suecica	YARRELL.

THE Blue-throated Robin is a rare accidental visitor, that breeds in the Arctic regions of Europe and Asia.

The nest is placed on the ground, among the larger herbage, on the sides of banks, and among low brushwood. It is well concealed, and is composed of roots, dried grass, and a little moss, the blossoms of the reed, leaves, small stalks, and is lined with finer moss, hair, and the beautiful down of the cotton-grass.

There are two broods, and the first is sometimes on the wing so early as the end of May. The male assists the female in the work of incubation.

The eggs, five to six in number, are of a greenish-blue colour.

BLUEBREAST.

REDSTART.

REDSTART

RED-TAIL—FIRE-TAIL—BRAN-TAIL—FIERY BRAN-TAIL.

PLATE CX.

Ruticilla phœnicurus,	.	.	.	MACGILLIVRAY.
Sylvia phœnicurus,	.	.	.	LATHAM. PENNANT.
Phœnicura ruticilla,	.	.	.	YARRELL.

THE nest of this summer migrant, which is more or less well concealed, and rather loosely constructed, is built of moss, dry grass, and leaves, and lined with hair and feathers. It is frequently placed in a hole in an old wall, under the eaves of a house, in a hollow or hole in a tree, or even between the branches of one, as also against a wall, if extraneous support is afforded. One has been known to have been placed in a watering-pot, others in flower-pots, and one in a hole in the ground. It is frequently placed close to or in the wall of a house, and that where persons are constantly passing, even within reach of the hand. Another has been known also placed on the ground under an inverted flower-pot; the hen bird successfully reared her brood, the flower-pot, which was at first unwittingly removed, having been replaced : the circumstance is related by the Rev. J. C. Atkinson in the *Zoologist.* Bishop Stanley mentions one he had known " built on the narrow space between the gudgeons or narrow upright irons

on which a garden door was hung; the bottom of the nest,
of course, resting on the iron hinge, which must have shaken
it every time the door was opened. Nevertheless, there she
sat, in spite of all the inconvenience and publicity, exposed
as she was to all who were constantly passing to and fro.
Another has been known in like manner to sit through the
din of three looms at work from five o'clock in the morning
until ten at night, within twelve feet of the nest. The same
situation, if the birds have been undisturbed, is frequently
resorted to from year to year. One pair have been known
to revisit the same garden for sixteen seasons in succession :
a pair resorted for four successive years to the ventilator
of a stable. The female is sedulously devoted to her eggs
or young, and will sometimes suffer herself to be touched
before flying off from the nest; if, however, they be
molested she will forsake it : both birds indeed are most
assiduous in their attentions to their brood, one or other of
them being to be seen in constant motion, conveying food
to them, or retiring in search of it. In one instance, the
male bird having been killed while the hen was sitting,
another partner joined the widow, and became foster-father
to the orphaned family." It has been known to lay its
eggs in the nest of a Titmouse.

The following was in the *Ipswich Journal* of June
11th, 1853 :—" In the gardens at Holbrook House, the
residence of Miss Reade, a little bird called the Redtail
has built a nest in an inverted flower-pot, six and a half
inches deep, and seven inches wide at the top. The hole
in the bottom, or rather the top as the pot stands, is one
and a half inches over, and through this the little bird
has carried the whole of the materials for its nest, which
is formed on the side of the pot. Six eggs were laid,

from which five young ones were hatched. The pot stands by the side of a gravel walk, at a spot where the family and gardener are continually passing."

The eggs, which are of a uniform light greenish-blue colour, are generally from five to six or seven in number, but as many as eight have been found. They are occasionally speckled with red spots.

One brood only is generally reared in the year.

BLACKSTART

BLACK REDSTART—BLACK RED-TAIL.

PLATE CXI.

Ruticilla tithys,	.	.	.	.	SCOPOLI.
Sylvia tithys .	.	.	.	.	JENYNS.
Phænicura tithys,	.	.	.	.	YARRELL. GOULD.

THE nest of this bird, which has rarely been known to breed in England, is rather large, is placed among the clefts of stones or rocks, and also in the holes of walls and ruins, the spires, towers, and higher parts of churches, and the roofs of houses. It is formed of grasses, straw, moss, wool, and the dry stalks and fibres of plants, and is lined more or less with hair or feathers. "It is," says Mr. W. R. Fisher, "formed of almost any material which is suitable and can be readily obtained. I have found it composed of grey worsted, taken from a loose ball which was lying in a garret."

The eggs, from five to six or seven in number, are usually of a very pure glossy white colour, sometimes with a faint tinge of blue or brown.

Two broods are reared in the year, the first being hatched by the beginning of May. The same situation is frequently returned to year after year.

BLACKSTART.

.

STONECHAT.

STONECHAT

PLATE CXII.

Pratincola rubicola,	. . . . .	LINNÆUS.
Saxicola rubicola, .	. . . . .	NEWTON.

THE Stonechats pair in March, and commence building before the end of that month.

The nest, which is large and loosely put together, and composed of moss, dry grass, and fibrous roots, or heath, lined with hair and feathers, and sometimes with wool, is placed among the grass or other herbage at the bottom of a furze or other bush, or in the bush itself, as also in heather, and even occasionally in some neighbouring hedge adjoining the open ground which the bird frequents. It is exceedingly difficult to find, on account of its situation in the middle of a cluster of whin bushes—such not admitting of the most easy access—the female also sitting very close, and, when off the nest, being very watchful, hopping quickly from bush to bush, and disappearing suddenly by retreat into cover.

The eggs, generally four to five or six in number, are of a pale greyish or greenish blue colour, the larger end minutely speckled with dull reddish brown. They are laid the middle or latter end of April, sometimes in the earlier part of that month, and have been known so late as the 12th of July.

Only one brood is usually reared in the season.

WHINCHAT.

WHINCHAT

GRASSCHAT—FURZECHAT.

PLATE CXIII.

Pratincola rubetra, ... DRESSER.
Saxicola rubetra, ... NEWTON.

THE Whinchat is a summer migrant which breeds generally over Great Britain.

The nest is placed in the lower part of a gorse bush, a few inches above the ground, where the thorns and stalks are dying off, so that the materials of the nest assimilate in appearance to the situation in which it is placed, and it is thus the rather screened from observation. Frequently it is placed in the grass at the foot of a thick furze bush. Where there are no gorse bushes, it is placed among rough grass in a pasture field, or in a meadow. Mr. Henry Stowe, of Emmanuel College, Cambridge, took one near Brackley in Northamptonshire, built so near the edge of a pond that the nest was quite wet. It is loosely built of stalks of grass and moss, and is lined with finer portions of the former; and occasionally some hair or leaves: it measures six inches across, and two and a half internally. It is very carefully concealed, and extremely difficult to find, as the bird approaches it stealthily.

The eggs are of a glossy bluish-green colour, some times with minute specks of dull reddish brown ; they are four to six in number, very rarely seven.

The young are hatched towards the end of May, one brood only being generally produced in the season.

CXIII

BLACK-THROATED WHEATEAR

BLACK-THROATED CHAT.

PLATE CXIII.*—FIGURE I.

Saxicola stapazina, ... Vieillot.

A SINGLE specimen of this species has been shot in Lancashire. It breeds in South Europe and North Africa.

The nest is a loose structure of stems of grass, &c., and is placed in holes and crevices in old walls and buildings.

The eggs are of a pale greenish tint, speckled with brown.

ISABELLINE WHEATEAR

ISABELLINE CHAT.

PLATE CXIII.*—FIGURE II.

Saxicola isabellina, ... RÜPPELL

ONE specimen of this species was found in Cumberland by the Rev. H. A. Macpherson in 1888. It is an inhabitant of East Africa and India, extending into China.

The nest, which is described by Heuglin as tolerably bulky and lined with soft grasses, is usually placed in the burrows of small animals.

The eggs resemble those of the Common Wheatear, being pale blue in colour.

DESERT WHEATEAR

DESERT CHAT.

PLATE CXIII.*—FIGURE III.

Saxicola deserti, ... RÜPPELL.

ONE specimen of this bird was shot in Scotland in 1880, and another in Yorkshire in 1885. It breeds in North-East Africa and the adjacent countries.

Its nest is placed in holes and crevices in rocks and walls, as also in the burrows of animals, and under bushes.

The eggs are of a greenish-blue colour, with liver-coloured spots on the larger end.

WHEATEAR

FALLOW-CHAT — WHITE-TAIL — STONE-CHACKER — CHACK-
BIRD—CLOD-HOPPER.

PLATE CXIV.—FIGURE I.

Saxiola œnanthe, ... Linnæus.

THIS regular visitant breeds on the open downs
throughout England.

Its nest, which is commenced the middle of May, is
sometimes well hid in the innermost recess of some crevice
among rocks, in an old wall, stone quarry, gravel-pit, or chalk-
pit, and frequently in a deserted rabbit-burrow, or the hollow
under some large clod, tuft, or stone. Mr. Hewitson has
known one in the bank of a river, in a hole deserted by a
Sand Martin. It is rudely constructed of loose fine dry
stalks of grass, and lined with rabbit's fur, hair, or feathers.

The eggs, usually from four to six in number, sometimes, though very rarely, seven, are of an elegant rather
elongated form, and of a uniform delicate pale blue colour,
deepest at the larger end usually, spotless, but sometimes
dotted with purple.

The young are abroad from the middle of May to June.
A second brood is usually produced in the season.

74

GRASSHOPPER WARBLER

CRICKET BIRD.

PLATE CXIV.—FIGURE II.

Locustella nævia, .	. . . . .	BODDAERT.
Sylvia locustella, .	. . . . .	NAUMANN.
Curruca locustella,	. . . . .	FLEMING.
Salicaria locustella,	. . . . .	SELBY.

THE nest, of a cup shape, is formed in a rather firm manner of reeds or grass, with sometimes a little moss, lined with finer portions of the same. It is difficult to find, owing to the careful habits of the bird, and is placed on the ground, and has been met with at the foot of a small bush by the roadside; it is completely hidden in the middle of some large tuft of fen grass, through which there is no apparent entrance but such as the bird threads for herself, creeping along like a mouse to and into it.

The eggs are from five or six to seven in number, of a pale-reddish white colour, freckled all over with specks of darker red; they seldom vary much.

Two broods are sometimes reared in the year.

SAVI'S WARBLER

PLATE CXV.

Locustella luscinioides,	.	.	.	.	.	SAVI.
Sylvia luscinioides, .	.	.	.	.	.	GOULD.
Salicaria luscinioides,	.	.	.	.	.	YARRELL.

THE nest of this rare summer visitant, which is placed on the ground, is formed of the leaves of the reed, wound round and interlaced, but without any other lining. It is begun the middle or towards the end of May, by which time, or early in June, the eggs are laid.   Both birds sit.

The eggs are of a whitish colour, minutely speckled nearly all over with pale red and light grey, in some the red, and in·others the grey, predominating.

SAVI'S WARBLER

SEDGE WARBLER

SEDGE BIRD—SEDGE WREN—REED FAUVETTE.

PLATE CXVI.

Acrocephalus phragmitis,	.	.	.	BECHSTEIN.
Sylvia phragmitis,	.	.	.	TEMMINCK.
Salicaria phragmitis,	.	.	.	YARRELL.
Calamoherpe phragmitis,	.	.	.	MACGILLIVRAY.

THE nest of the Sedge Warbler is sometimes placed at about two, and never at a greater height than three or four feet from the ground, on a stump of a willow or alder tree, but generally among the tall grass or flags that grow along the side of the river or pool. In the north of England, the Rev. H. A. Macpherson says that it shows a much greater predilection for nesting in hedgerows at a distance from water than is common in the south. The nest is made of moss, stalks of grass, and other smaller plants, lined with finer parts of the same and hair: it is rather large, and but loosely put together.

The eggs, generally five to six in number, are of a pale yellowish-brown colour, marked with light brown and dull grey. They are usually closely freckled all over, and often streaked at the large end with dark hair-lines: they vary considerably. Mr. Heysham mentions a nest which contained three quite white. Sometimes they are uniform dull yellow: they are laid early in May.

REED WARBLER

PLATE CXVII.

Acrocephalus streperus,	.	.	VIEILLOT.
Sylvia arundinacea,	.	.	NAUMANN.
Motacilla arundinacea,	.	.	MONTAGU.
Salicaria arundinacea,	.	.	GOULD. YARRELL.

THIS bird is common in the south, though rare in the north of England.

Its nest is a very artistic piece of work, and is generally placed between three, four, or five stems of the common reed that grow near to one another, at a height commonly of about three feet above the water, but has been known as much as nine or ten feet from the ground. To these the self-taught architect fastens the nest, twining and interlacing the materials of which it is composed round and round the reeds at intervals, until the whole is firmly fixed—not so firmly, however, but that the reeds may be easily slipped out without injuring the structure. It is formed of dried grass, moss, long stalks, lichens, and wool, and is lined with the blossom of the reed. It generally consists of two parts, a loose foundation of the first-named materials, and the actual nest, which is composed almost exclusively of the last-named. This upper part can sometimes be detached from the lower, as if from a socket, the whole being narrow

and deep to secure the eggs when the reeds are so swayed down, that the frail fabric, the bird all the while sitting in it, is often brought close to the very water's edge. The depth outside is from about three to five inches, and the inside about three, by about three in width at the top and two at the bottom. The nest, however, is not invariably placed among reeds; it is at times found in a blackthorn, whitethorn, willow, or among the clustering branches of an osier bed. Mr. Sweet met with one in the low part of a poplar tree, and Mr. Bolton another in a hazel bush. When destroyed by floods, these birds have been known to build repeatedly. Mr. James Dalton, of Worcester College, Oxford, has taken one from a box tree, near the piece of water which is there so great an ornament, and Mr. N. Rowe, of the same College, has found one in a lilac tree.

The eggs, usually four, but sometimes five in number, are of a dull greenish-white colour, spotted and freckled with darker greyish-green and light brown. In some instances the spots are almost black, in others inclining to a brownish green; occasionally the egg is marked with one or two little black lines at the broad end. The arrangement of the spots is endless—some varieties are equally marked all over; in some the spots are in a ring round the broad end; in others the base is covered; some are but slightly marked; others are completely clouded over; one rare variety has been seen almost white, faintly mottled with pale grey blots; some quite white have been known. They are frequently not laid until after the beginning of June.

The young are hatched in July, and quit the nest before they can fly, making their way about the stalks of the reeds with their parents.

AQUATIC WARBLER

PLATE CXVII.*

Acrocephalus aquaticus, ... GMELIN.

AN accidental visitor. Breeds in temperate Europe and North Africa.

Its nest, which is found in sedges and water plants, is similar to that of our Sedge Warbler, being made of moss, grass, roots, and neatly lined with hair and leaves, and is built on or near the ground.

The eggs, four to five in number, are pale yellowish brown, clouded with darker brown.

NIGHTINGALE.

NIGHTINGALE

PLATE CXVIII.

Daulias luscinia, . .	LINNÆUS.
Sylvia luscinia, . . .	NAUMANN.
Philomela luscinia, . .	YARRELL. MACGILLIVRAY.

THE nest of the Nightingale, which is almost always
placed on the ground, in some natural hollow, amongst
the roots of a tree, on a bank, or at the foot of a
hedgerow, though sometimes two or three feet from the
surface, is very loosely put together, and is formed of
various materials, such as dried stalks of grasses, and leaves,
small fibrous roots, and bits of bark, lined with a few hairs
and the finer portions of the grass. It is about five inches
and a half in external diameter, by about three internally,
and about three and a half deep.

The eggs, of a regular oval form, are of a uniform
glossy dull olive-brown colour. They are sometimes tinged
with greyish blue, especially at the smaller end; some are
greenish, others brownish green; some are paler, mottled
with olive or reddish brown. They are four or five to six
in number. They are laid in May, one brood only being
reared in the year. The young, which are hatched in June,
often leave the nest before they are able to fly.

Mr. Meyer observes: "The attachment of this species

to its young, and its grief at their loss, have been noticed by many writers, ancient and modern. Our friend, the Rev. E. J. Moor, sends us, on this subject, a memorandum from his journal: 'One evening while I was at college,' he says, 'happening to drink tea with the late Rev. J. Lambert, fellow of Trinity College, he told me the following facts illustrative of Virgil's extreme accuracy in describing natural objects. We had been speaking of those well-known lovely lines in the fourth Georgic on the Nightingale's lamentation for the loss of her young, when Mr. Lambert told me that, riding once through one of the toll-gates near Cambridge, he observed the keeper of the gate and his wife, who were aged persons, apparently much dejected. Upon inquiring into the cause of their uneasiness, the man assured Mr. Lambert that he and his wife had both been made very unhappy by a Nightingale, which had built in their garden, and had the day before been robbed of its young. This loss she had been deploring in such a melancholy strain all the night, as not only to deprive him and his wife of sleep, but also to leave them in the morning full of sorrow; from which they had evidently not recovered when Mr. Lambert saw them.'"

CXIX

THRUSH NIGHTINGALE

NORTHERN NIGHTINGALE.

PLATE CXIX.

Daulias philomela,	.	.	.	DRESSER.
Sylvia turdoides,	.	.	.	MEYER.
Philomela turdoides,	.	.	.	BLYTH. GOULD.

THIS species is of very doubtful occurrence in Great Britain, but is common in the north of Europe during the period of migration.

The nest is built in small thickets, but most frequently in low and damp situations.

The eggs are of a brownish olive-colour, stained with deep brown.

GREAT REED WARBLER

PLATE CXX.

Acrocephalus turdoides,	.	.	.	MEYER.
Salicaria turdoides,	.	.	.	YARRELL.

THE Great Reed Warbler is common on the Continent, but is very rarely seen in England, although its large size and chattering song would be sure to attract attention.

It is not known to have nested in this country. Its nest, which is found in the reed-beds of temperate Europe, is cup-shaped, some five inches deep, and formed of dry grass and the blossoms and tops of reeds. The whole is woven into and suspended from several upright reed stems.

The eggs, four to six in number, are pale greenish blue, blotched and speckled with ash colour. The bird only rears one brood during the season, and in September migrates to Africa. This bird has unfortunately been described under other names. In the early edition of Yarrell it was known as the Thrush-like Warbler, and by some others it has been named the Great Sedge Warbler, but it is more correctly termed the Great Reed Warbler.

GREAT SEDGE WARBLER

CXX*

RUFOUS WARBLER

PLATE CXX.*—FIGURE I.

Ædon galactotes,	.	.	.	.	NEWTON.
Salicaria galactotes,	.	.	.	.	YARRELL.

THE Rufous Warbler is a rare accidental visitant to England, breeding in the south of Europe and north of Africa.

The nest is placed in a bush, or on the ground. It is built of the twigs of trees, and lined with feathers and hair.

The eggs are of a pale grey-white colour, spotted and streaked, speckled with shades of ashy brown; they are from three to five in number.

MARSH WARBLER

PLATE CXX.*—FIGURE II.

Acrocephalus palustris, ... BECHSTEIN.

A VERY rare visitant, breeding in temperate Europe and Asia.

The nest is placed in swampy thickets, never over-hanging the water, though often close to it ; but not on ground itself. It is made of grass stalks, and lined with hair.

The eggs, five to seven in number, are of a whitish ground colour, blotted and spotted with olive brown and violet grey. There are two types according to Seebohm, some having the ground colour pale greenish blue.

BARRED WARBLER

PLATE CXX.*—FIGURE III.

Sylvia nisoria, . . . Bechstein.

THIS bird has only occurred some three or four times in England.

The nest is built in a bush, near the ground, or on a branch of a tree, at no great elevation generally, but sometimes as high up as twenty-five feet.

The eggs, generally five in number, are buff-white marbled with grey.

BLACKCAP

BLACKCAP WARBLER—MOCK NIGHTINGALE.

PLATE CXXI.

Sylvia atricapilla,	. .	PENNANT. JENYNS. LINNÆUS.
Motacilla atricapilla,	. .	MONTAGU. BEWICK.
Curruca atricapilla,	. .	GOULD.

THE nest, built about the end of May or the beginning of June, is commonly placed in a bramble or other bush, sometimes in a honeysuckle, a raspberry, or currant tree, about two or three feet or rather more from the ground; a privet-hedge being often selected. · It is made of dry grass and small fibrous roots, with occasionally a little moss and hair—the latter as a lining, and the outer parts cemented together with spiders' webs and wool. It is strong and tolerably compact, though slight. Anything like meddling with it, or intruding upon it, is jealously watched, and the smallest disturbance causes the nest to be forsaken. Several in fact are frequently abandoned, either from apprehension or caprice, before they have been finished. Prof. Alfred Newton mentions in the *Zoologist* his having found a nest on the 11th of March 1845, which contained an egg at that early date.

The eggs, usually four or five in number, sometimes six, are of a light brown and grey, with a few spots and

BLACKCAP.

streaks of olive, dusky, and dark brown. They are subject
to considerable variation. Some are marbled with deeper
shades of reddish brown; white ones have at times been
found. They vary a good deal in size and shape.

Both birds sit on the eggs, but the female naturally
the most. The male often sings while so engaged, and
thus not unfrequently betrays the position of the nest.
The female, when sitting, is occasionally fed by her partner.
The young leave the nest rather soon, roosting with their
parents on the adjoining boughs.

ORPHEAN WARBLER

PLATE CXXII.

Sylvia orphea, .	. .	TEMMINCK.
Sylvia grisea, .	. .	VIEILLOT.
Curruca orphea, .	. .	GOULD.

AN accidental visitor to this country. A specimen bird was shot on the 6th of July 1848, in a small plantation near Wetherby, in the West Riding of Yorkshire, and preserved by Mr. Graham of York. It was a female, and appeared to have been sitting the same summer: the male bird was also observed with it for a considerable time previously. An account of this interesting occurrence was published in the *Zoologist.*

The Orphean Warbler builds sometimes in low bushes, such as tamarisks, and in young cork trees, often in company with others of the same species. The nest is composed of small twigs, leaves, and long grass, interwoven with horse-hair, and lined with the down of cotton-grass.

The eggs are four or five in number, greyish white, irregularly marked with brown spots of various shades, chiefly at the larger end.

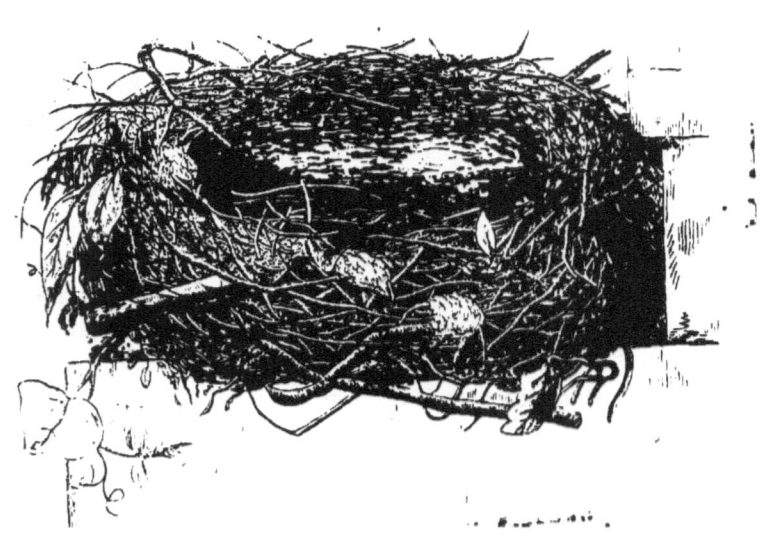

GARDEN WARBLER.

GARDEN WARBLER

GREATER PETTYCHAPS.

PLATE CXXIII.

Sylvia hortensis, .	.	LATHAM. BECHSTEIN.
Curruca hortensis,	.	SELBY.

THE nest of this well-known migrant is made of the bents of straws and small roots, mixed sometimes with a small quantity of moss, and lined with a little wool or horse-hair, and fine fibres of plants. It is generally placed between the branches of some low blackthorn, white-thorn, or other bush not far from the ground, as also at times on the ground among the taller wild plants. It is rather loosely constructed. One is said to have been found in an open field among some tares, and it has been found among peas or gooseberry bushes in gardens. Mr. Jesse mentions his having found one three times in succession among some ivy growing against a wall. It is not very carefully concealed.

The eggs are four or five in number, of a dull yellowish grey, or pale brown, spotted and blotted with darker markings of the latter colour.

Both male and female are believed to take their turn on the nest. Only one brood, as a rule, is commonly reared in the season.

WHITETHROAT

COMMON WHITETHROAT—MUGGY—NETTLE-CREEPER.

PLATE CXXIV.

Sylvia cinerea,	.	PENNANT. JENYNS.
Motacilla sylvia, .	.	MONTAGU. BEWICK.
Curruca sylvia, .	.	FLEMING.
Curruca cinerea, .		GOULD.

THE nest of this common visitor is loosely compacted.
It is placed near the ground, or not more than two
or three feet above it, in a low hedge, or sometimes in
a bramble, furze, sloe, wild rose, or other bush, as also
frequently among nettles or other tall weeds or herbaceous
plants on the ground, or beside a bank; Mr. Jesse mentions
one which built in a vine close to a window. It is composed chiefly of dried stalks of grasses, though other plants
are occasionally used, and lined with a good deal of hair of
various kinds, with which it is often, though not always,
thickly woven on the inside, giving it accordingly more or
less consistency. The same situation is frequently resorted
to year after year. A trifling disturbance will cause the
owner to desert before the eggs are laid, but the reverse
is the case afterwards. Not much care is taken in its concealment. The young quit the nest early, even before
they are fully able to fly, if alarmed for their safety. Two
broods are reared in the season; in the south of Scotland,

WHITETHROAT.

however, the first nest is seldom completed before the end of May. The bird occasionally builds close to a public road, or in the immediate vicinity also of a dwelling-house.

The eggs, four, five, or six in number, are of a greenish-white ground colour, with spots and speckles of grey and brownish grey. Some are more of a stone-coloured ground.

LESSER WHITETHROAT

PLATE CXXV.

Sylvia sylviella,	.	.	PENNANT. MONTAGU.
Sylvia curruca,	.	.	TEMMINCK.
Motacilla curruca,	.	.	LINNÆUS.
Motacilla sylviella,	.	.	BEWICK.
Curruca sylviella,	.	.	FLEMING.
Curruca garrula,	.	.	GOULD. MACGILLIVRAY.

THE nest of this regular migrant, which is begun about three weeks after the arrival of the birds, is of a slight construction, and is made of dry grass and a little wool or moss, lined, but rarely, with small fibres, roots, and hairs; it is rather loosely interwoven, and is bound together with spiders' webs and such like materials. It is sometimes placed among the herbage on a bank, as well as in the lower part of a hedge, or in some low shrub—a nut tree, gooseberry bush, blackthorn, broom, woodbine, and among briers and brambles, generally at a height, in the latter, of about four or five feet from the ground, but sometimes as high as even ten.

The eggs, five to six, exceedingly pretty, are of a white or creamy white colour, spotted, most numerously at the larger end, and sometimes in the way of a zone, with small dots and patches of brown, olive brown, and light grey.

Incubation lasts from twelve to fourteen days, com-

mencing about the 20th of May. Two, and sometimes even possibly three, broods are reared in the season.

The young birds in their nestling plumage nearly resemble the old ones, but the colour of the head and the back is more uniform.

WOOD WARBLER

WOOD WREN — GREEN WREN — LARGER WILLOW WREN —
YELLOW WILLOW WREN.

PLATE CXXVI.

Phylloscopus sibilatrix,	. . .	BECHSTEIN.
Sylvia sylvicola, . .	. . .	YARRELL.
Sylvia sibilatrix,	. . .	SELBY.

THE nest of this species, which is domed or half domed, and of an oval shape, is almost always placed on the ground, among herbage in woods, the entrance being through a small hole in the side. It is made of grasses, leaves, and moss, cleverly but not·thickly interwoven, lined with horse-hair, but not with feathers. It is well concealed, and is usually to be found on the side of some sloping wooded banks. Mr. Sweet says that he has often found the nest on the stump of a tree.

The eggs, six, or more commonly seven in number, are of a white ground colour, thickly spotted and speckled all over with dark purplish brown and violet grey, forming a mass at the larger end. Some are much less deeply marked than others.

Like the eggs of all the family, says Booth, they lose their beauty soon after incubation commences. Those seen in the cabinet of the collector bear but a faint resemblance to the appearance they presented when fresh laid.

96

WOOD WARBLER.

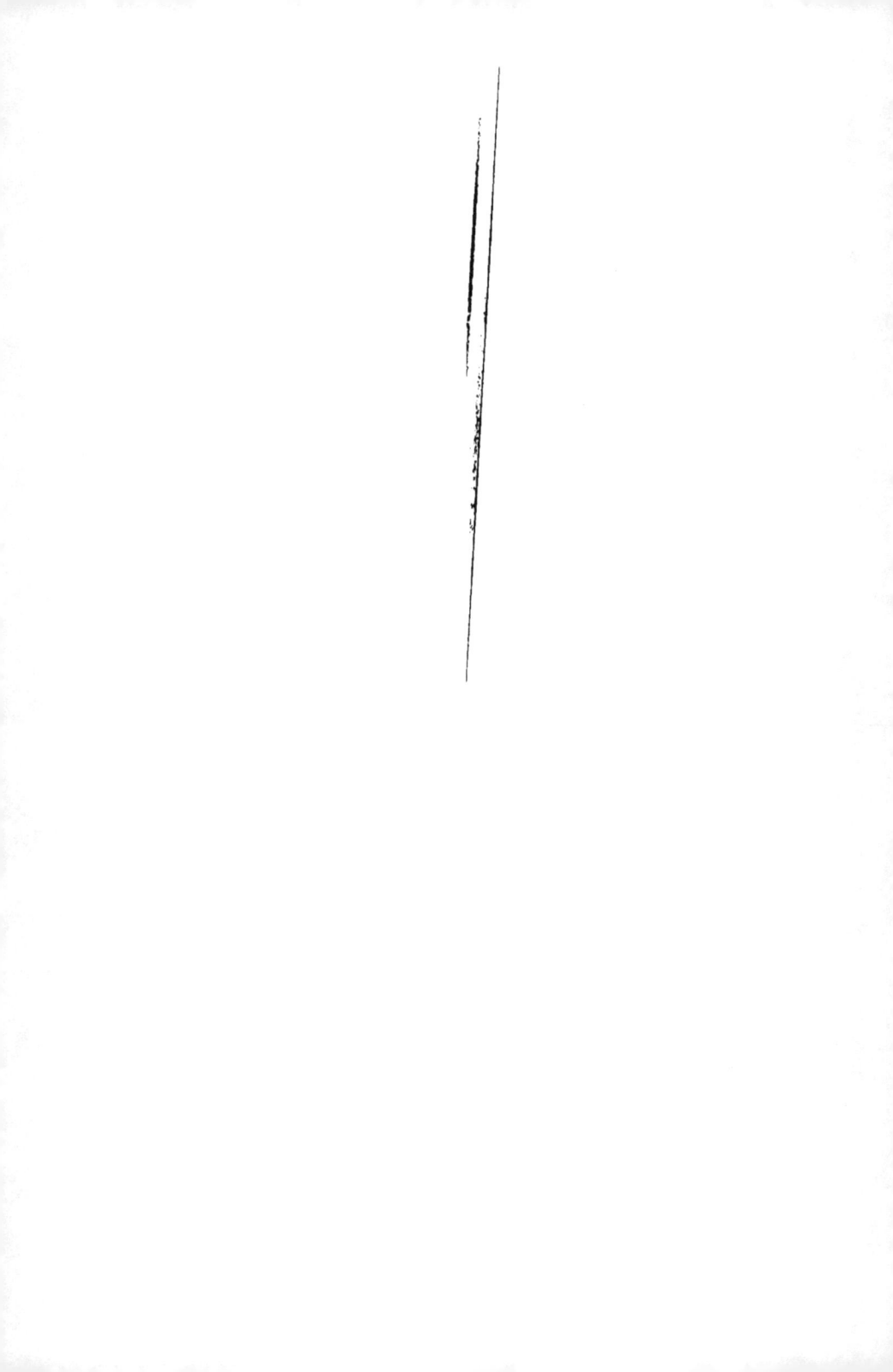

WILLOW WARBLER.

WILLOW WARBLER

YELLOW WARBLER—WILLOW WREN—HUCK-MUCK.

PLATE CXXVII.

Phylloscopus trochilus,	.	NEWTON.
Sylvia trochilus,	. .	PENNANT. SELBY. JENYNS.
Motacilla trochilus, .	.	MONTAGU. LINNÆUS.

THE nest of this common summer visitor is very large for the size of the bird, of an oval but rather flat shape, though it varies in form, according to the situation in which it is placed, being domed or semi-domed, and is built of moss, leaves, or fern and dry grass, a hollow being left in the side for the ingress of the bird. It is lined with a profusion of feathers, and with hair, the former being the innermost, and is pretty firmly compacted. It is placed on the ground, generally in woods, or among the long grass, brushwood, or weeds on the bank of some wooded hedge by the outside of a wood, or the edge of a pathway or open place in such. One has been met with in the ivy on a wall, and another in a field, several yards from the fence. Mr. James Croome informs me of one placed two yards from a fence, in long grass, which having been destroyed, a second was built, and a third, the second having been also accidentally destroyed. The nest is rather carefully concealed.

The eggs, of a rotund form, but varying much in size

and marks, are from six to eight in number, and are white, with numerous small specks of pale rusty red ; some are less thoroughly spotted, and some most marked at the larger end, while others are only sparingly dotted ; they are a little polished : pure white ones have been met with. The female bird sits very close upon them, and the male feeds her in the nest, taking her place in the course of the day, while she searches for food.

The young are hatched the end of May or beginning of June, and are fledged about the middle or end of that month, or the beginning of July. A second brood is generally reared during the season, and is abroad by the beginning of August.

MELODIOUS WILLOW WARBLER.

MELODIOUS WILLOW WARBLER

ICTERINE WARBLER—MELODIOUS WILLOW WREN.

PLATE CXXVIII.

Hypolais icterina,	.	VIEILLOT.
Sylvia hippolais, .	.	TEMMINCK.

THIS bird has only occurred thrice in Great Britain. Mr. Gould says that this species builds on trees, as well as at times in shrubs in gardens. The nest is formed of dry grass, wool, thistle-down, and lichens, lined with hair.

The eggs are four to five in number, of a reddish-white or dull rose-pink colour, blotted and speckled with spots and dots of darker red or purplish brown.

CHIFF CHAFF

PLATE CXXIX.

Phylloscopus rufus,	.	. .	BECHSTEIN.
Sylvia rufa,	.	. .	TEMMINCK.
Motacilla hippolais,	.	. .	LINNÆUS.
Sylvia hippolais,	.	. .	YARRELL.

THE nest of this extremely common migrant is arched over, skilfully constructed of various indiscriminate materials, according to the situation it is placed in, fern, moss, leaves, grasses, bark, the shells of chrysalides, wool, and the down of flowers, with abundance of feathers and a few hairs for lining for the whole of the interior; it is arched over more than half-way; if the roofing be removed, even three or four times, the bird will often renew it. It is placed on the ground, generally, but not always, in the immediate neighbourhood of trees, or on a hedge bank, or near a brook, or on the moss-clad stump of a tree, beneath the shelter of the trailing boughs of some bramble, furze, or other bush, or clod of earth. Mr. Henry Doubleday has found one at a height of two feet from the ground, in some fern; and Mr. Hewitson mentions another, which was built in some ivy against a garden wall, at a like elevation. Occasionally the nest is placed in a row of peas, or a bed of ground-growing wild plants. I have seen one on the top of a wall in

CHIFF CHAFF.

Londesborough Park, at a height of six feet from the ground; and on being disturbed, the bird built a little farther on in some ivy against the side of the wall, about four feet up.

The eggs, usually six in number, are more than ordinarily rounded at the larger end: they do not vary much, and are of a white ground colour, with very small dots and spots of pale red or purple brown, chiefly at the thicker end, which they sometimes surround in the way of a zone or belt. Mr. Neville Wood saw a nest which contained five eggs of the usual colour, and the sixth pure white. The shell is but little polished. The eggs are laid towards the middle or end of May, and the young birds are fledged about the middle of June: they quit the nest early.

Incubation lasts thirteen days, and the male ocasionally relieves the female at her post. Two broods are sometimes reared in the season.

DARTFORD WARBLER

FURZE WREN.

PLATE CXXX.

Sylvia provincialis,	. . . .	SEEBOHM.
Sylvia undata,	. . . .	BODDAERT.
Motacilla provincialis,	. . . .	GMELIN.
Melizophilus provincialis,	. . .	MACGILLIVRAY.

THE nest of this bird, which is now known to be a resident in the furze districts of the south of England, is slight in its make, is placed in a furze bush, to the stems of which it is attached, at a height of about two feet from the ground. It is built of dry stalks and gorse grass, mixed with bits of the gorse; the materials, though in reality firmly compacted, are apparently but loosely put together, and have a slight interweaving of wool.

Two broods appear to be reared in the year, the second nest being more flimsy than the first. Montagu found the nest and eggs after the middle of July, the earlier brood being hatched early in May.

The eggs, four to five in number, are of a whitish-grey ground colour, slightly tinged with green, speckled all over with olive-brown and ash-colour; near the larger end the markings are more run together, and form a sort of zone.

WREN

COMMON WREN—KITTY WREN—JIMPO.

PLATE CXXXI.

Troglodytes parvulus, ...		KOCH.
Sylvia troglodytes, ...		LINNÆUS.
Troglodytes vulgaris, ...		TEMMINCK.
Troglodytes europæus, ...		CUVIER.

THE nest, very large in size in proportion to the bird, and ordinarily of a spherical shape, domed over, but flattened on the side next the substance against which it is placed, varies much both in form and substance, according to the nature of the locality which furnishes the materials and a *locus standi* for it. It is commenced early in the spring, even so soon as the end of the month of March, the birds pairing in February. The nests are made of fern and moss, grass, small roots, twigs, and hay, closely resembling in most cases the materials amongst which they are placed; some are lined with hair or feathers, and others not. The nest is firmly put together, especially about the orifice, which is strengthened with small twigs or moss, and nearly closed by the feathers inside. It is in thickness about one inch to two inches, and about three inches wide within by about four in depth, and outside about five wide by six deep. At times they are found on the ground, and also in banks, as well as against trees, even so high up as twenty feet, also under the eaves of the thatch of a building, in holes in walls, the sides of stacks, among piles of wood or faggots, or the bare roots of trees, and under the projection at the top of the bank of a river; one has been known to be placed in an old bonnet fixed up among some peas to frighten the birds. Mr. Hewitson mentions one built against a clover stack.

Other situations for nests are the tops of honeysuckle and raspberry bushes, in the latter case the nest being made of the leaves of the tree; in fir trees, trelliswork, granaries, the branches of wall-fruit trees, and lofts, use being made occasionally of the holes previously tenanted by Sparrows and Starlings. One has been known to be built inside that of a Swallow, and another in the old nest of a Thrush: one, again, in the newly-finished nest of a Martin, another on a branch of a yew tree among the foliage, and another in one of the hatches in the river at Winchester. Mr. Jesse relates a curious anecodote of a Wren's nest, the owner of which being disturbed by some children watching it, blocked up the original entrance, and opened out a new one on the other side. In the garden of Nunburnholme Rectory one was built, in 1854, in the middle of a low quickset hedge, near the top, close to the walk. It was composed nearly entirely of dry leaves.

The male feeds the female while sitting. Two broods are produced in the season. The least disturbance will cause the nest to be forsaken and a new one built; even then, if the eggs or the young be once handled, this species will some-times desert them. This, or some interruption of the ordinary course of laying, may be the cause of tenantless nests of the Wren being so frequently found; it is, however, said that a forsaken nest will sometimes be again returned to. Thus several nests of the same year are often found near together, the work of one and the same pair of birds; and other nests, in the making of which both birds assist, are not very unfrequently put together in the autumn, and in these the birds shelter themselves in the winter, possibly as being of the newest, and therefore the best, construction, and made too late in the year for a further brood: these nests seldom, if ever, contain any feathers. The young are said to return to lodge in the nest for some time after being fledged.

The eggs are usually from six to eight in number, but generally not more than eight, though as many as a dozen, or even fourteen, have been found, of a white colour, sprinkled all over with small spots of dark red, which are most numerous at the obtuse end; some are quite white: the shell is very thin and polished.

GOLDCREST

GOLDEN-CRESTED KINGLET—GOLDEN-CRESTED WREN—
GOLDEN-CROWNED WREN.

PLATE CXXXII.

Regulus cristatus,	.	.	LINNÆUS.
Regulus auricapillus,	.	.	SELBY. JENYNS.
Regulus vulgaris,		.	GOULD.
Motacilla regulus,	.	.	MONTAGU.
Sylvia regulus,	.	.	PENNANT. MONTAGU.

THESE birds begin to pair even by the end of February, and Mr. Selby has known the young birds fully fledged so early as the third week in April, the nest being built in March. They build a second time.

The nest is placed underneath and generally near the end of the branch of a fir, or occasionally on an oak, cypress, holly, yew, cedar, or other tree, as also not very unfrequently in a laurustinus or other bush, and, though very rarely, in a hedge. It is attached by the moss and lichens of which it is composed being interwoven with the smaller shoots. It is built with willow down, moss, cocoons, spiders' webs, wool, lichens, grasses, and a few hairs. It measures about three inches and a half in diameter inside, and is deep and of a spherical shape, the orifice being almost always in the upper part. It

GOLDCREST.

closely assimilates in colour to the branch to which it is
fixed. In a fir it is mostly composed of moss, and, in a
thorn tree, of lichens. It is sometimes placed near the
top of the tree, and in other instances only two or three
feet from the ground. These birds have been known to
steal the materials from the nests of Chaffinches to make
their own; one was noticed to do so most slily, watching
its opportunity, but on the Chaffinch detecting and chasing
it, it did not repeat the theft. The nest is frequently lined
with feathers, and is altogether a singularly elegant piece
of architecture; the feathers are so placed as to project
inward. Two nests have been found on one branch. Mr.
Hewitson says: "It is sometimes placed upon the upper surface
of the branch; and I have also seen it, but rarely, placed
against the trunk of the tree upon the base of a diverging
branch, and at an elevation of from twelve to twenty feet
above the ground." He also mentions, in the *Zoologist*,
his having once met with the nest in a low juniper bush,
very little more than a foot from the ground. Mr. James
Croome writes of one he found in the stump of a thorn
bush about four feet from the ground, and another in a
bush a few feet from the hedge at a height of about six
feet. Deserted nests of this species are frequently to be
met with, but the reason is not known.

The eggs are four, five, six, or seven, to eight, or even
ten or eleven in number; they are of a very pale reddish
or brownish white, the larger end being darker coloured;
some have been known pure white, sparingly spotted with
reddish brown here and there. They are smaller than those
of any other British bird, and are sometimes almost of a
globular shape. The young are fed by both the parents.
Two broods are reared in the year, and the second is less

numerous than the first. Eggs, fresh laid, have been met with in May and June, while young birds have been known fully fledged by the third week in April. The same nest has also been known to have been used twice in the same season, two broods being hatched and reared.

CXXXIII

FIRECREST

FIRE-CRESTED KINGLET—FIRE-CROWNED KINGLET—
FIRE-CRESTED WREN.

PLATE CXXXIII.—FIGURES I. AND II.

Regulus ignicapillus,	.	JENYNS. MACGILLIVRAY.
Sylvia ignicapilla,	.	TEMMINCK.

THE nest of this accidental visitor is similar to that of the Goldcrest, being built of moss, wool, and a few grasses, filled with spiders' webs, studded with lichens, and lined with fur and feathers. It is suspended from the branch of a fir or other tree.

The eggs are from seven to ten in number, similar to those of the Goldcrest, but much redder in the ground colour and dots, as shown in the second figure on the plate.

YELLOW-BROWED WARBLER

YELLOW-BROWED WILLOW WREN—DALMATIAN REGULUS.

PLATE CXXXIII.—FIGURE III.

Phylloscopus superciliosus, . GMELIN.
Regulus modestus, . . GOULD.

THE nest of this accidental visitor is of slighter construction than that of the Goldcrest, made of grass and moss, sparingly lined with finer grass, reindeer's hair, or feathers, and a few spiders' egg-bags on the outside. It is placed on the ground, by preference near the edge of a wood, or at the root of some small bush or tree.

The eggs are six in number, pure white, mottled over with reddish brown, especially about the larger end; some much marked, and others only minutely spotted.

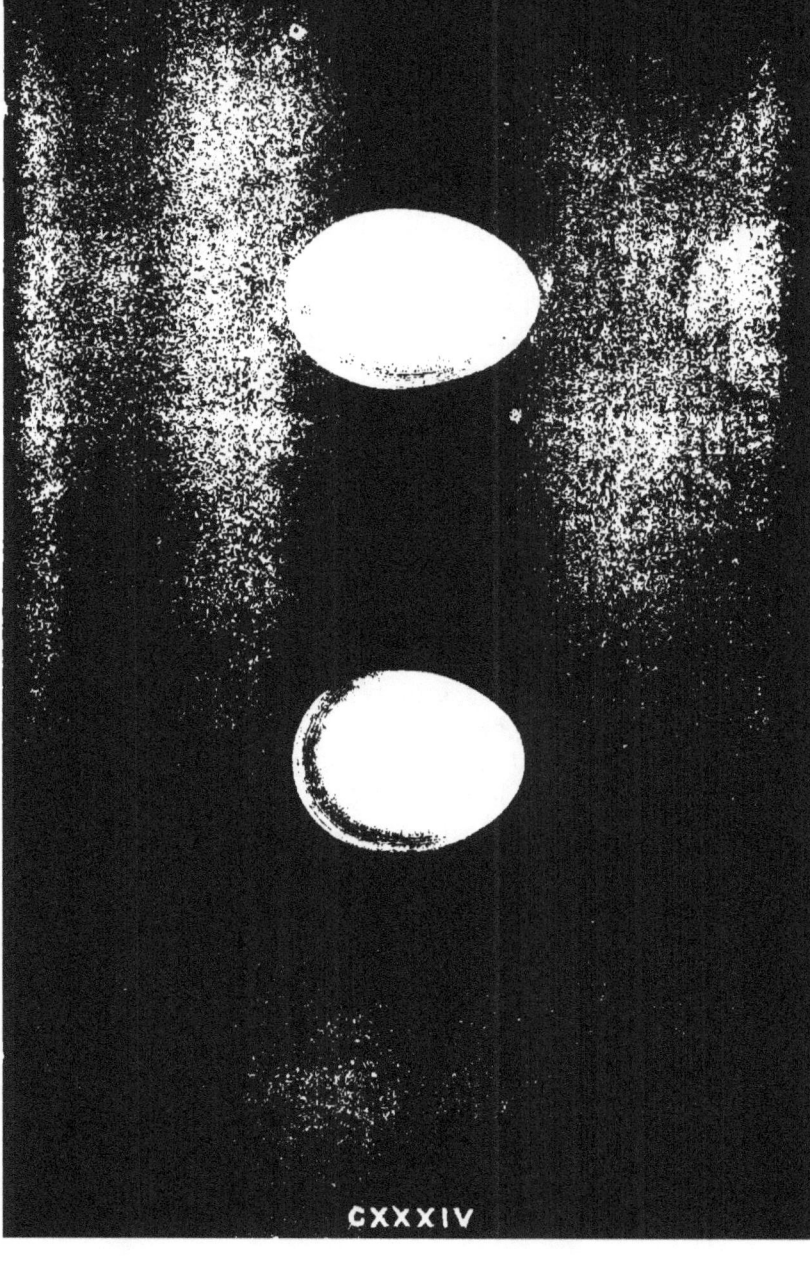

CXXXIV

WOOD PIGEON.

RING DOVE—CUSHAT—QUEEST.

PLATE CXXXIV.—FIGURE I.

Columba palumbus, LINNÆUS.

THE nest of the Wood Pigeon is wide and shallow, placed in almost any kind of tree, and frequently in thick ivy on cliffs or old walls; it is little more than a rude platform of a few crossed sticks and twigs, the largest as the foundation, so thinly laid together that the eggs or young may often be seen from below. It is often built in woods and plantations, but not unfrequently also in single trees, even those that are close to houses, roads, and lanes—the oak and the beech, the fir, or any other suitable one—or even in ivy against a wall, rock, or tree, or in a thick bush or shrub in a garden, or an isolated thorn, even in the thick part, so that in flying out in a hurry, if alarmed, many of the loosely-attached feathers are pulled out. One pair built in a spruce fir not ten yards from a garden gate, where they were constantly liable to disturbance by the ringing of the bell, and the passing in and out of the members of the family. Another pair dwelt two years in succession close to a window by a frequented walk, and this though a cat destroyed the young. Many are now built in the trees in the parks and squares of London.

The eggs are always two in number, pure white, and of a rounded oval form; two and sometimes three broods are produced in the season, but the third may possibly be only the consequence of a previous one having been destroyed; the eggs are hatched in eighteen days. The young are fed from the bills of the parent birds with soft curdy food when in the nest. The male and female both take their turns in hatching the eggs and in feeding the young, the former sitting from six to eight hours—from nine or ten in the morning to about three or four in the afternoon.

The first brood are abroad by the beginning of May; the second in the end of July. Macgillivray has known the young unfledged in October, and a pair with down tips to the feathers on the 26th of that month. Mr. Hewitson, too, has recorded young as late as the middle of September.

STOCK DOVE

PLATE CXXXIV.—FIGURE II.

Columba anas, ... LINNÆUS.

THE Stock Dove is rather an early breeder, usually laying in April.

The nest, which is flat and shallow—a mere layer of twigs slightly put together—is often placed on the ground in an old deserted rabbit burrow, on the bare sand or earth, a few sticks being occasionally used; and in such places under furze and other bushes, where the surface is hollowed; also, ordinarily, in any suitable holes in trees, or clefts, and in pollard tops and matted ivy. The same hole is resorted to again. A second and sometimes a third brood is reared in the year. Booth in his Rough Notes says: " While in quest of the young of a Tawny Owl in a large wood adjoining Balcombe Forest, we alighted on a brood of juvenile Stock Doves in a squirrel's drey on the limbs of an antiquated oak standing in a dense thicket." Incubation lasts eighteen days, and in about a month the young are able to fly. The parents are very careful of the eggs, and will even sit on them till they are taken off with the hand.

The eggs, like those of the Doves and Pigeons in general,

are two in number; sometimes more are found, but then two birds have laid in the same nest: they are pure creamy white, smaller than those of the Queest, and somewhat pointed at the smaller end and of an oval shape. They have been known to be laid as late as the 2nd of October.

CXXXV

ROCK DOVE

ROCKIER.

PLATE CXXXV.—FIGURE 1.

Columba livia, . . . SELBY. JENYNS. GOULD.

THE Rock Dove builds in companies in rocky cliffs on the coast, many often in the same cavern. The nest is composed of sticks and dry stalks, with blades of grass and other plants, laid together without much care. The bed is fresh made without much trouble for a new brood as soon as the former has been sent at large. The first eggs are laid about or towards the middle of April, and the latest the latter end of August; the young are seen about the end of September.

The eggs are white, and two in number; while the hen is sitting, the cock bird remains at night close to the nest. The young birds are fledged in about three weeks, and are fed from the crops of their parents for some days after they are able to fly.

TURTLE DOVE

PLATE CXXXV.—FIGURE II.

Columba turtur, .	.	LINNÆUS. LATHAM.
Turtur auritus, .	.	SEEBOHM.
Turtur cornunius,	. .	SAUNDERS.

THE Turtle, unlike our other Doves, is a summer visitant common only in the southern counties of England.

Its nest, built in woods and hedges, is frail and carelessly constructed of a few twigs and sticks, and is placed in trees or thick bushes at no great height from the ground—some ten or twenty feet—but well hidden among the foliage. It is, however, itself so slight, that the eggs may be seen through it.

The eggs are two in number, and glossy white, of a narrow, oval, and rather pointed form. They are laid late in May or early in June, and are hatched in eighteen days. The female sits on the young, if the weather be cold, both night and day. Two broods are sometimes produced in the year.

PASSENGER PIGEON.

PASSENGER PIGEON

PLATE CXXXVI.

Columba migratoria,	.	. .	FLEMING. YARRELL.
Ectopistes migratoria,	.	. .	SELBY.

A VERY rare straggler from North America, of which five or six examples only have been shot in the British Islands.

The nest, which is placed in trees, and is only a layer of a few sticks, is put together in a single day. The young are hatched in eighteen days; both male and female assisting in making the nest, in the work of incubation, and in feeding the young.

The eggs, two in number, and not one as frequently stated, are pure white.

PHEASANT

COMMON PHEASANT.

PLATE CXXXVII.

Phasianus colchicus, ... LINNÆUS.

THE nest, a very slight fabrication of a few leaves, is made upon the ground, sometimes in the open fields, but more commonly in woods and plantations, among underwood, under fallen or felled boughs and branches of trees, in long grass, and in hedgerows: a few feathers sometimes become detached from the bird, and are found among the eggs.

The eggs are begun to be laid in April and May; incubation lasts twenty-four days. The eggs usually are from ten to fourteen in number, smooth, and of a light olivebrown colour, minutely dotted all over. Some are greyish white tinged with green. The hen sits on the chicks for some time after they are hatched, and they keep with her till they begin to moult to the full plumage. When half grown they roost with her in the trees. It would appear that two hens will sometimes lay in one and the same nest, and also that that of the Partridge will occasionally be made use of, even if it already contain eggs, the Pheasant expelling their proper owner, and hatching them with her own, and bringing up the young.

CXXXVII

CAPERCAILLIE

WOOD GROUSE.

PLATE CXXXVIII.

Tetrao urogallus,	.	.	. .	LINNÆUS.
Urogallus major,	.	.	. .	BRISSON.

THE Capercaillie, which, after becoming extinct, has been reintroduced into this country, usually nests in May, and the young are hatched early in June.

The nest, composed of grasses and leaves, is made upon the ground, in long grass or heath, under the shelter of a tree, or bramble, or other bush. One has been known at a good height from the ground, in a pine tree, in an old nest of a Falcon.

The eggs are from half-a-dozen to a dozen in number, of a pale reddish-yellow brown, spotted all over with two shades of orange brown. Incubation is said to last about a month, the hen alone sitting, the male keeping in the neighbourhood. If danger approaches, she runs off a little way, but returns again as soon as she can with safety. The young leave the nest soon after they are hatched, and keep with the mother bird till towards the approach of winter.

The account of the reintroduction of this bird has been admirably given by Mr. J. A. Harvie-Brown in his work "The Capercaillie in Scotland."

BLACK GROUSE

BLACK GAME—BLACK COCK.

PLATE CXXXIX.

Tetrao tetrix, ... LINNÆUS.

THE nest of the Black Grouse is usually placed not far from water, or in a marshy spot, among heath, or in newly made plantations, and sometimes in hedge-rows, generally under the shelter of some low bush, or among high grass in some hollow, and is composed inartificially, but rather neatly, of grass and a few twigs laid together.

In the "Game Birds and Wildfowl" of Mr. Beverley R. Morris,* the author says, speaking of the time after the hen birds have commenced sitting: "They are deserted by the cock birds, who again assemble in small parties, and seek the secluded and quiet thickets, among which they chiefly remain till they have completed their moult. They are, during this seclusion, particularly timid and shy. The female has thus the whole charge of hatching and bringing up the young birds. . . . The packs of male birds are sometimes very numerous, often amounting to from fifty to seventy birds. The females also in autumn are occasionally

---

* "British Game Birds and Wildfowl," by Beverley R. Morris, M.D. Fourth Edition. London, J. C. Nimmo.

found in packs, but in much smaller numbers, generally under twenty."

The eggs are from five or six to ten in number, of a pale yellowish red or yellowish white colour, irregularly spotted and dotted with reddish brown. They are laid in May. Soon after the young birds are hatched they are taken to the higher parts of the moorland, and will generally be found amongst rank and coarse herbage on boggy ground, the food of the young birds chiefly consisting of the seeds of the rushes.

RED GROUSE

GOR-COCK—MOOR-COCK—MOOR-FOWL—MUIR-FOWL.

PLATE CXL

Lagopus scoticus,	. . . . .	LATHAM.
Tetrao scoticus,	. . . . .	SEEBOHM.

T HE Red Grouse pairs early in the spring, eggs being
often found in sheltered ground as early as March.
A nest with fifteen eggs was found on the 25th of March,
1835, on Shap Fell, Westmoreland. The female usually
begins to lay in March or April; she sits very close, and
may be even taken off her eggs.

The nest, which is made in a depression in the ground
usually under the shelter of a tuft of heather, is very scanty;
it is made of twigs of heather and grass, with occasionally
a few of the bird's own feathers.

The eggs are usually eight to ten or even more in
number, of different shades of ground colour—reddish white,
brownish yellow, yellowish grey, or yellowish white, thickly
clouded, blotted, and dotted with rich red or brown: they
are of a regular oval form.

While the young are hatching, the hen utters an occasional chuckle. The Heath Poults leave the nest shortly

RED GROUSE

after they are hatched, and are soon able to fly; they keep together till the end of autumn, unless dispersed by shooters: they are attended by both the parents. At the beginning of the season they lie close, but gradually become more wild as they are disturbed.

PTARMIGAN

PLATE CXLI.

Lagopus vulgaris,	. . .	FLEMING.
Lagopus mutus, .	.	SELBY. GOULD.
Tetrao lagopus, .	.	LINNÆUS.

THE Ptarmigan pairs early in the spring, and the eggs are begun to be laid in May, and are hatched by the beginning of July. The hen alone brings up the brood.

The nest, if any be formed, for sometimes the bare earth is laid upon, is composed of a small portion of heather or grass, placed in some slight hollow under a rock, stone, or plant, and is very difficult to be detected, "for," says Sir William Jardine, "the female, on perceiving a person approach, generally leaves it, and is only discovered by her motion over the rocks, or her low clucking cry." The male on the first sign of danger has flown off, and she thus follows him, the young dispersing in all directions, hiding themselves and laying still under any stones, tufts, or bushes. Meyer says: "It is reported that the male Ptarmigan behaves very remarkably during the time when the female sits on her eggs, and that under these circumstances he will sit immovable in one spot for hours together, even on the approach of danger; and when stationed thus near the nest he has been known to remain there, looking around on the landscape quite unmoved. As soon as the young are hatched, both

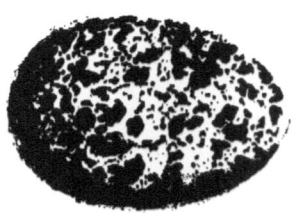

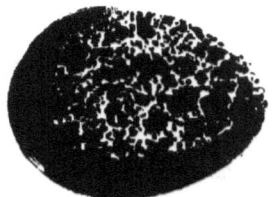

parents become alert and busy, and towards autumn more careful, and finally very shy in the winter. If the weather is fine and sunny in winter, they are all again slow to move." But the male, it would appear, leaves the rearing of the young to the hen bird, rejoining them all again later in the season, and then several families pack together.

The eggs, from eight to ten in number, of a regular oval form, are of a white, yellowish white, greenish white, or reddish colour, blotted and spotted with rich chocolate brown, and the ground colour varies greatly from dirty white to rich brownish buff.

SAND GROUSE

PLATE CXLI.*

Syrrhaptes paradoxus,   PALLAS.

A T the last migration of this singular species in 1888, it nested repeatedly in Great Britain, and in some cases the young were hatched and were afterwards figured.

It does not construct a nest at all, but deposits its eggs on the sand, sometimes without even making a hole. The eggs are of a regular elliptical shape, with a dirty yellowish-grey ground, marked with reddish and brownish spots and streaks.

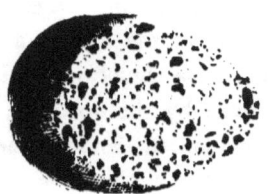

PARTRIDGE

COMMON PARTRIDGE.

PLATE CXLII.—FIGURE I.

Perdix cinerea,	. . . . .	LATHAM.
Tetrao perdix,	. . . . .	LINNÆUS.

THE Partridge begins to pair very early, even so soon as the beginning of February. At pairing time there are often fierce combats between the male birds.

The nest is only a few straws placed in a mere hollow scratched in the earth, under the shelter perhaps of some tuft, generally in open grass and other fields, among peas, corn, weeds, or herbage, at the foot of a tree or bush or by a post, but at times in a small plantation, among shrubs, under a hedgerow, even by the roadside, and on the moors in the vicinity of cultivated land; sometimes in holes of decayed trees, as much as three or four feet from the ground, and even in the thatch on the top of hay-stacks; I have been told of a nest placed in this situation, the brood hatched, and safely reared. Another I have heard of under the post of a hand-gate which was turned whenever passengers went backwards and forwards through it. A brace of Partridges have been known, their own nest having been destroyed, to take up with the nest and eggs of a pair of Pheasants, the hen of which had been killed, on the estate of of Colonel Burgoyne, in Essex. The hen bird alone sits, the male keeping watch, and when the young are hatched he joins the covey, and protects and feeds them with the dam.

The eggs, which are of a pale olive-brown colour without markings, are laid towards the end of May or the beginning of June; pale blue or whitish varieties are not unfrequent: they are usually ten or twelve in number, but sometimes as many as fifteen, eighteen, or even twenty. Twenty-two eggs are recorded to have been found in one nest, and thirty-one in another, two hen birds having occupied the same one; and in the former instance the cock bird gathered half of the united family under his wings, the pair sitting side by side. In two other instances thirty-three eggs are recorded as having been found in one nest, but there is little doubt that they were contributed by more than one bird. In one of these twenty-three young were hatched and went off, and four of the other eggs had live birds in them. The young leave the nest almost as soon as they are hatched. Incubation lasts about twenty-one days, beginning usually in June, about the 20th, as has been stated, but no doubt generally earlier, especially in the south, though often later—in 1874, in February, in Scotland. A Partridge's nest was found at Thistlewood, Cumberland, containing seventeen Partridge's eggs and six common Hen's eggs. The Partridge and the Hen were sitting together upon the nest.

"It is a curious fact," says Mr. Jesse, "that when young Partridges are hatched and have left the nest, the two portions of each shell will be found placed the one within the other. I believe that this is invariably the case. This is doubtless done by the chicks themselves in their last successful effort to escape from prison." Only one brood is reared in the year, unless indeed the first nest be destroyed, but in these cases the eggs are fewer, and the young are said to be less strong.

RED-LEGGED PARTRIDGE

FRENCH PARTRIDGE.

PLATE CXLII.—FIGURE II.

Caccabis rufa,	.	.	. .	LINNÆUS.
Perdix rufa,	.	.	. .	MONTAGU.

THE nest of the Red-legged Partridge is made of grass and a few feathers of the bird itself, and is placed on the ground among corn, grass, clover, or growing crops.

Mr. Jesse says that a clergyman in the county of Norfolk found the nest in the thatch of a hay-rick, and informed him that such is no unfrequent occurrence. Other similar instances are mentioned.

The eggs are usually from ten to fifteen in number: as many as eighteen have been sometimes found. They are of a reddish yellow-white colour, spotted and speckled with reddish brown. The young leave the nest soon after being hatched. The male takes no part in the incubation of the eggs, and leaves the care of the brood to their mother till they are half grown, when he returns to them, and continues with them till the following spring.

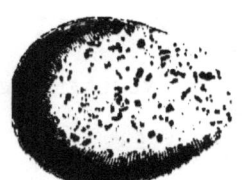

BARBARY PARTRIDGE

PLATE CXLIII.—FIGURE II.

Caccabis petrosa,	. . . . . .	DRESSER.
Perdix petrosa,	. . . . . .	LATHAM.

THESE birds, which are very rare stragglers to England, build in barren places and among desert mountains, among low bushes on the ground.

The eggs are as many as fifteen, of a dull yellowish colour, thickly dotted with greenish-olive spots.

VIRGINIAN PARTRIDGE

VIRGINIAN COLIN.

PLATE CXLIII—FIGURE I.

Ortyx virginianus,	. . .	SAUNDERS. HARTING.
Perdix virginiana,	. . .	LATHAM. JENYNS.

THIS bird, which has been repeatedly introduced in large numbers from North America, has never become established in this country.

The nest, placed under or in some thick tuft of grass that shelters and conceals it, is described as well covered with a hood, an opening being left at one side for entrance, and is composed of leaves and fine dry grass, both birds assisting in its fabrication.

The eggs, from ten or twelve to fifteen or even twenty-four in number, this latter quantity the joint produce in all probability of two birds laying in the same nest, are pure white, without any spots, and broad at one end, but pointed at the other.

The hen bird performs the task of incubation, and the whole family keep together till the following spring. The young leave the nest at once on being hatched, and are conducted forth by the female in search of food, and from time to time are sheltered under her wings, collected together by a twittering cry. Should danger appear to threaten, she displays extreme anxiety, boldly attacking an intruder, or using every artifice and stratagem to draw him away, feigning lameness, "throwing herself in the path, fluttering along, and beating the ground with her wings, as if sorely wounded, uttering at the same time certain peculiar notes of alarm well understood by the young, which dive separately among the grass, and secrete themselves till the danger is over; and the parent having decoyed the pursuer to a safe distance, returns, by a circuitous route, to collect and lead them off." She shows the greatest assiduity and the most sedulous and unremitting attention to their further care. Wilson mentions a curious anecdote of some young ones which had been hatched under a hen, and which, "when abandoned by her, associated with the cows, which they regularly followed to the fields, returned with them when they came home in the evening, stood by them while they were milked, and again accompanied them to the pasture. These remained during the winter, lodging in the stable, but as soon as spring came they disappeared."

QUAIL

COMMON QUAIL.

PLATE CXLIV.

Coturnix communis, ...	BONNATERRE.
Perdix coturnix	LATHAM. JENYNS.
Tetrao coturnix, ...	LINNÆUS.

THIS migrant is thinly distributed in England in the summer.

For a nest the female scrapes out a small hollow in the ground, into which she collects a few bits of dry grass, straw, clover, and such like. It is generally placed in the open amongst growing crops or herbage. She alone sits, and very closely, on the eggs, but the male assists her in the care of the young.

The eggs are yellowish white, orange-coloured white, or greenish, blotted or speckled with brown. They vary much in number, from six to twelve, or even, it is said, twenty, though generally ten; a bevy of ten birds has been known to be reared. Incubation lasts about three weeks. Two broods are sometimes reared in the season The eggs are not laid till June, or even July. The young follow the dam as soon as they are hatched.

ANDALUSIAN QUAIL

ANDALUSIAN HEMIPODE—ANDALUSIAN TURNIX—
THREE-TOED QUAIL.

PLATE CXLV.

Turnix oylvatica, ...		DESFONTAINES.
Hemipodius tachydromus, . . .		YARRELL.
Turnix tachydroma, ...		MEYER.

THREE examples of this bird only have been obtained in England.

The birds nest in North Africa and the south of Europe, on the ground, under dense shelter, and, from the skulking habits of the birds, the nest is difficult to find.

The eggs, four in number, are described as of a dirty white colour, blotched with purplish grey and brown.

GREAT BUSTARD

PLATE CXLVI.

Otis tarda, ... PENNANT. MONTAGU.

THIS species, formerly common in England, has been gradually exterminated as far as this country is concerned.

The eggs are laid in a hollow scraped on the bare earth. " It is said that the Great Bustard will forsake her nest, if only once driven from it by apprehension of danger ; but when the eggs are laid, and still more when the young are produced, it is only repeated meddling with them that will induce the parents to forsake them."

The eggs, two to three in number, are of an olive greenish-brown colour, blotted with pale ferruginous and ash-coloured spots.

GREAT BUSTARD.

LITTLE BUSTARD

LESSER BUSTARD.

PLATE CXLVII.—FIGURE I.

Otis tetrax, ... Linnæus.

THIS species is only a rare winter straggler, not nesting in this country.

The nest, a mere hollow in the ground, is made with a few dry grasses, and placed under the shelter of any sufficiently high herbage that will conceal the bird.

The eggs are from three to five in number, olive brown in colour, sometimes varied with patches of a darker shade of brown.

MACQUEEN'S BUSTARD

PLATE CXLVII.—FIGURE II.

Otis macqueenti, ... GRAY.

MACQUEEN'S Bustard is accidental in Europe, being an Indian species. It has only once occurred in England.

The egg is of a dull olive-brown colour, with some irregular blots over it.

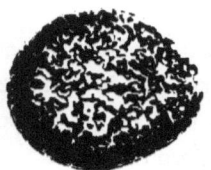

PRATINCOLE

PLATE CXLVIII—FIGURES I. AND II.

Glareola pratincola, ... Dresser.
Hirundo pratincola, ... Linnæus.

THIS very rare visitor to Great Britain breeds chiefly in North Africa, Asia Minor, and the south of Europe— wintering in South Africa.

According to Mr. Seebohm, who is familiar with its habits, "The birds of this species do not make any nest, but lay their eggs upon the bare ground, seldom, if ever, taking the trouble to scratch a hollow or to collect what dry grass or seaweed may be at hand. They seem studiously to avoid coarse grass or rank herbage, and prefer to lay their eggs on the dried mud, sheltered only by the straggling plants of Salsola, which grow all over the lowest and wettest parts of the islands. The number of eggs was usually two, occasionally three, and only in one instance four; probably the latter clutch was the production of two females."

The eggs are very oval in form, exceedingly fragile, the ground colour being buff or grey, spotted with streaks and blotches of black or purplish brown.

139

COURSER

PLATE CXLVIII.—FIGURE III.

Cursorius gallicus, ... GMELIN.

Cursorius europæus, ... LATHAM.

THE eggs of Coursers, which are laid in the barest parts of North Africa, Asia Minor, Persia, and India, are two or three in number: they are very handsomely mottled over with reddish brown, on a lighter ground colour.

GREAT PLOVER

STONE CURLEW—NORFOLK PLOVER—THICK-KNEE.

PLATE CXLIX.

Œdicnemus crepitans,	.	.	.	NAUMANN. SEEBOHM.
Œdicnemus scolopax,	.	.	.	DRESSER.
Charadrius crepitans,	.	.	.	MONTAGU. BEWICK.
Charadrius ædicnemus,	.	.	.	LINNÆUS. GMELIN.

THE Stone Curlew breeds locally in England, on the chalk downs and open heaths. The eggs are laid on the bare earth, among weather-worn stones. The male appears to sit as well as the female. The young are led about by the female almost as soon as hatched : at first the old birds take great care of them.

The eggs are pale clay brown, blotted, spotted, and streaked with darker brown, assimilating closely in appearance to the grey flints that surround them, thus being very difficult to detect. They are generally two in number, but Mr. Allan Hume has frequently taken three in India.

Only one brood is reared in the year, but if the first eggs are removed, the birds will nest again even as late as September.

GOLDEN PLOVER

WHISTLING PLOVER—YELLOW PLOVER—GREEN PLOVER

PLATE CL.

Charadrius pluvialis, ...	LINNÆUS.
Charadrius auratus, ...	NAUMANN.

THE Golden Plover commences its nest in Great Britain about the middle of May.

The nest is a very simple structure, being merely a few stems of grass and fibres laid together in some small hollow of the ground, only just large enough to contain them; what there is, is made the end of April or beginning of May.

The eggs, four in number, are large in proportion to the size of the birds. They are usually of a yellowish stone-colour, blotted and spotted with brownish black. They are placed quatrefoil—with the small ends pointed together inwards.

The young quit the nest as soon as hatched, and follow their parents till able to fly and support themselves, which is in the course of a month or five weeks.

Mr. Booth describes the Golden Plovers as breeding in considerable numbers on the Grouse moors of many of the northern counties of the Highlands; he states, "I have come across their haunts repeatedly in Perthshire, Ross-shire, Sutherland, and Caithness, as well as in the Western Islands."

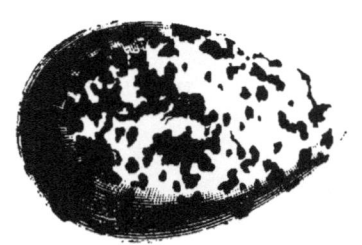

DOTTEREL

PLATE CLI.

Eudromias morinellus, ...	DRESSER.
Charadrius morinellus, ...	YARRELL.

THIS summer migrant breeds in Scotland and the north of England.

Any small hollow in the ground serves for a nest, and it is generally near some stone or rock; a few lichens, moss, or short grass make its "mossy bed." The male assists the female in the work of incubation, which lasts apparently for eighteen or twenty days. The hen bird sits very close, and if disturbed only runs a few yards off.

The eggs never exceed three in number. They are laid from the end of May and the beginning of June to the end of June and even the beginning of July.

Their ground colour varies from greyish buff to yellowish olive, blotted and spotted with brownish black.

RINGED DOTTEREL

RINGED PLOVER—SAND LARK—SAND LAVEROCK.

PLATE CLII.

Ægialitis hiaticula, ... DRESSER.

Charadrius hiaticula, ... NAUMANN.

THE nest of this common species is but a slight
natural hollow amongst small gravel, or on sand, frequently under the shelter of some tall grass; it is generally
placed on a bank by the beach, just above high-water
mark, but occasionally in sandy places farther inland, as
much, Sir William Jardine says, as ten, or from that to
fifteen or twenty miles; in some instances on the banks
that line the coast, or even over them in an adjoining field.
The Ringed Plover is common on warrens in Norfolk and
Suffolk, and also in the Fens of Bottisham and Swaffham, in
Cambridgeshire.

The eggs are four in number, pear-shaped, and of a
greenish grey, pale buff, or cream colour, spotted and
streaked with bluish grey and black, or blackish brown.
The male and female both sit on them and appear much
attached to each other, as well as very careful of their
eggs and young.

The birds lay generally by the middle of April, pro-
ducing two broods in the season, recently hatched young
being often found in the first week in August.

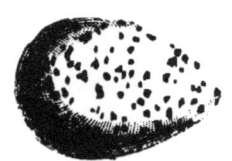

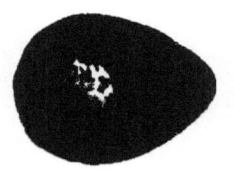

LITTLE RINGED DOTTEREL.
KENTISH DOTTEREL.

LITTLE RINGED DOTTEREL

LITTLE RING DOTTRELE—LITTLE RINGED PLOVER.

PLATE CLIII.—FIGURE I.

Charadrius minor, .	. . .	MEYER.
Ægialitis curonica, .	. . .	GMELIN.

THIS bird is one of the rarest of our occasional visitants.

The eggs, four in number, are of a pale stone colour, with numerous small spots of bluish ash, reddish brown, and dark brown. They are laid in a hollow in the sand, without any lining. The young are hatched in sixteen or seventeen days, and at once begin to run about, hiding themselves instinctively with much cleverness if endangered.

KENTISH DOTTEREL

KENTISH PLOVER.

PLATE CLIII.—FIGURE II.

Charadrius cantianus,	.	.	.	LATHAM. JARDINE
Charadrius albifrons,	.	.	.	MEYER.
Charadrius littoralis,	.	.	.	BECHSTEIN.
Ægialitis cantiana,	.	.	.	DRESSER.

THE nest of this species is placed on the shingle or sand, any slight depression serving as a receptacle for the eggs; a few blades of grass or withered weeds may perchance afford a scanty lining.

The eggs are three or four in number—Mr. Gould, erroneously, says five; they are of a yellowish colour, finely and much marked all over, but chiefly at and about the centre and base, with dark blackish brown.

GREY PLOVER

GREY SANDPIPER.

PLATE CLIV.

Squatarola cinerea,	.	.	FLEMING. SELBY. GOULD
Vanellus melanogaster,	.	.	TEMMINCK.
Tringa squatarola,	.	.	PENNANT. MONTAGU.
Charadrius helveticus,	.	.	SEEBOHM.

THE eggs, four in number, resemble those of the Golden Plover, but are browner.

The Grey Plover is a tolerably common visitant to our coast. On its two seasons of passage it breeds in the tundras of Siberia, its nest being a slight hollow on the moorland.

PEEWIT

LAPWING.

PLATE CLV.—FIGURE II.

Vanellus cristatus,	.	.	.	.	FLEMING. SELBY.
Charadrius vanellus,		.	.	.	NAUMANN.

THE nest of this common resident is a small and slight depression in the soil, with the addition sometimes of a few bits of grass, heath, or rushes; the footprint of a cow or horse being frequently utilised.

The eggs, which are usually four in number, are very delicate eating, and sold in immense numbers for the purpose. They are so disposed in their narrow bed as to take up the least amount of room, the pointed ends laid inwards, towards the centre of the nest. They vary to an extraordinary degree, though those in each nest are generally very much alike: some are blotted nearly all over with deep shades of brown. In general they are of a deep dull green colour, blotted and irregularly marked with brownish black. They are wide at one end and taper to the other, as is the case with birds of this class. They are hatched in fifteen or sixteen days.

One brood only is generally reared in the year, but if the first clutch of eggs be removed others will be laid.

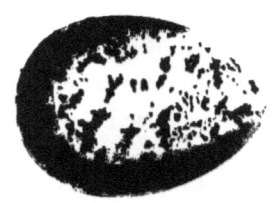

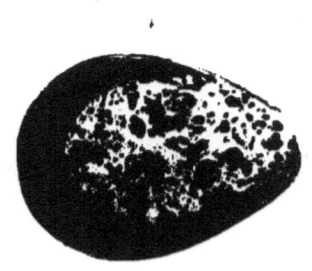

TURNSTONE.
PEEWIT.

TURNSTONE

COMMON TURNSTONE.

PLATE CLV.—FIGURE I.

Strepsilas interpres,	.	.	.	FLEMING. SELBY.
Tringa interpres,	.	.	.	LINNÆUS.
Tringa morinella,	.	.	.	LINNÆUS.
Charadrius interpres,	.	.	.	SEEBOHM.

THESE birds lay their eggs in the north of Europe and Greenland, on sandy and rocky coasts, where a stunted vegetation obtains. They appear to have no tie to any previously tenanted situation, but choose a new residence, if it suit them, year after year. The nest is sometimes placed under the shelter of a stone, rock, plant, or other break in the surface, and at other times on the mere rock, sand, or shingle. It is but some trifling hollow, natural or scraped out for the purpose, lined, perhaps, with a few dry blades of grass, or leaves.

The eggs, four in number, vary much in colour and markings, some being of a green olive ground, and others of a brown olive colour; some much and others only a little spotted, principally about the obtuse end, with dark grey, olive brown, and black, or reddish-brown of two shades. They are cleverly concealed.

Mr. Hewitson says that all the eggs of this species that

he met with, in his visit to the coast of Norway, were suffused with a beautiful purplish tint seen in those of few other species.

One brood only is reared in the season, both birds taking part in the incubation.

KILLDEER PLOVER

PLATE CLV.*—FIGURE I.

Ægialitis vocifera, ... LINNÆUS.

A SINGLE example of this American species has been recorded as having been killed in England, and a second in the Scilly Islands.

The nest is, as with many others of such birds, a mere hollow in the ground, lined only, so far as it is lined at all, with a few bits of dry grass, among which are found small pieces of shells, but these, I should imagine, introduced accidentally, and not intentionally; sometimes it is made on a heap of seaweed.

The eggs, generally four in number, are of a yellowish stone-colour, or pale buff spotted and variously marked with blackish brown.

SOCIABLE PLOVER

PLATE CLV.*—FIGURE II.

Vanellus gregarius, ... PALLAS.

A SINGLE example of this Eastern species has been recorded as having been obtained in Great Britain. Little is known respecting its nidification; eggs have only been obtained from the Moravian colonists at Sarepta, on the Volga.

The egg is of a clear yellowish brown colour, more or less mottled over with dark reddish brown, chiefly at the larger end.

SANDERLING

CURWILLET—TOWILLY.

PLATE CLV.*—FIGURE III

Arenaria calidris,	. . . . .	GOULD
Arenaris vulgaris,	. . . . .	STEPHENS.
Calidris arenaria,	. . . . .	LINNÆUS.
Charadrius calidris,	. . . . .	PENNANT.
Charadrius rubidus,	. . . . .	GMELIN.

THE nest is said to be placed in marshy places, and formed in a rude manner of grass. Col. W. H. Fielden, of the Polar Expedition, found one on a gravel ridge, at a height of several hundred feet above the sea, in a hollow in a low willow bush, lined with a few leaves and catkins. One brood only is reared in the year.

The eggs are described as being four in number, buffish olive, according to Seebohm, and thickly spotted with olive brown. Those figured by Mr. Fielden are of a bright yellowish brown, speckled over with spots of a darker shade.

This bird, which is common in winter only in Great Britain, breeds in the Arctic regions.

OYSTER-CATCHER

PIED OYSTER-CATCHER—SEA PIE—OLIVE.

PLATE CLVI.

Hæmatopus ostralegus, . . . PENNANT. MONTAGU.

THE nest is placed among gravel or stones, or among grass near the sea bank, in situations above high-water mark, where these materials of building are at hand, and the bird seems to be especially partial to a mixture of broken shells, which it carefully collects together and places in a slight hollow in the ground, using considerable care in their disposition. Several nests appear to be made sometimes, before one gives perfect satisfaction ; many nests are also placed in contiguity to each other, intermixed too, it may be, with those of other aquatic birds. Some have been met with on the top of isolated rocks, at a height of from ten to fifteen feet from the ground. In lieu of shells, small pieces of stone or gravel are selected. Incubation lasts about three weeks. "In many parts of the Highlands," says Mr. Booth, "they rear their young in a potato or oat field, the female sitting plainly in view until the crops get up sufficiently to afford concealment. While travelling by the Highland railway from Dunkeld towards Aberfeldy or Blair-Athol, I often watched several birds sitting on their eggs in the fields near the line."

The eggs are three and occasionally four in number, of a yellowish stone-colour, spotted with grey, dark brown, and brownish black. They have been found in April, May, June, and July, so that although one brood only seems to be usually reared in the year, if the eggs are taken it would appear that others may be laid. The eggs are disposed with their small ends inwards.

END OF VOL. II.

Printed by BALLANTYNE, HANSON & CO.
Edinburgh and London